WINEMAKING BASICS

C. S. Ough, DSc, MS

SOME ADVANCE REVIEWS

"Dr. Ough's book is carefully organized and written in a very readable style. It thoroughly covers each topic so as to be a valuable resource for winemakers, investors, or administrators in the winemaking industry, and for home winemakers. The references are up to date. Overall, this will be a very practical and useful winemaking manual."

John Almy, PhD
Professor, Department of Chemistry
California State University, Stanislaus
Turlock, California

"An extraordinary range of facts regarding winemaking. To the serious home winemaker, it is a must for his library. Ough manages to update all the traditional concepts and embellish with recent scientific thinking. For professional and experienced winemakers, this book should be a ready reference. In a field as romantic and hearsay as wine, this book stands out for its objective and factual presentation."

Edmund A. Rossi, Jr.
Consultant, Heublein Wines
Madera, California

"This book on the current state of the science and art of winemaking will be welcome by all those who make wine. All of the author's experience and the best current practices have been integrated into an up-to-date text that will repay careful study by winemakers everywhere."

Maynard A. Amerine, PhD
Professor Emeritus of Enology
University of California at Davis

NOTES FOR PROFESSIONAL LIBRARIANS AND LIBRARY USERS

This is an original book title published by Food Products Press, an imprint of The Haworth Press, Inc. Unless otherwise noted in specific chapters with attribution, materials in this book have not been previously published elsewhere in any format or language.

CONSERVATION AND PRESERVATION NOTES

The paper used in this publication meets the minimum requirements of American National Standard for Information Sciences—Permanence of Paper for Printed Material, ANSI Z39.48-1984.

Winemaking Basics

FOOD PRODUCTS PRESS
An Imprint of The Haworth Press, Inc.
Robert E. Gough, PhD, Senior Editor

New, Recent, and Forthcoming Titles:

Vintage Wine Book: A Practical Guide to the History of Wine, Winemaking, Classification, and Selection, Second Edition by Sommelier Executive Council

Winemaking Basics by C. S. Ough

Statistical Methods for Food and Agriculture by Filmore E. Bender, Larry W. Douglass, and Amihud Kramer

World Food and You by Nan Unklesbay

Understanding the Japanese Food and Agrimarket: A Multifaceted Opportunity edited by A. Desmond O'Rourke

Highbush Blueberry Management by Robert E. Gough

Winemaking Basics

Cornelius S. Ough, DSc, MS

Food Products Press
An Imprint of The Haworth Press, Inc.
New York • London • Norwood (Australia)

Published by

Food Products Press, an imprint of The Haworth Press, Inc., 10 Alice Street, Binghamton, NY 13904-1580.

Library of Congress Cataloging-in-Publication Data

Ough, C. S.
Winemaking Basics / C.S. Ough.
p. cm.
Includes bibliographical references and index.
ISBN 1-56022-005-8 (acid free paper). – ISBN 1-56022-006-6 (pbk.)
1. Wine and wine making. I. Title.
TP548.074 1991
663'.2 – dc20

91-2253
CIP

This effort is fully dedicated to Anne – my wife and true friend of many years – whose courage and spirit were a great inspiration and driving force, not only for this book but also for my life.

ABOUT THE AUTHOR

Cornelius S. Ough, DSc, MS, has been a faculty member of the Department of Viticulture and Enology at the University of California at Davis for 41 years. He is currently Professor of Enology, and past Department Chair, and an Enologist in the Experimental Station. The author of over 200 technical journal articles on winemaking, Dr. Ough is also the co-author of two texts, one on winemaking and one on wine analysis which was recently awarded the Office International de la Vigne et du Vin outstanding book prize. He has traveled to Israel and Brazil for the Food and Agriculture Organization of the United Nations and has spent two sabbatical leaves in South Africa and one in Australia.

Dr. Ough, a consultant to the Technical Committee of the Wine Institute, served as president of the American Society for Enologists and Viticulturists. He is a member of the American Chemical Society (Agriculture Chemistry), the American Society for Enologists and Viticulturists, New York Academy of Science, and the South African Society for Enology and Viticulture.

CONTENTS

Preface xi

Introduction 1

Chapter 1: Grapes and Concentrate 5

Vineyard 5
Grape Maturity 16
Wine Grape Varieties 23
Harvesting 39
Buying Grapes (Home Winemaker) 43
Raisins (Home Winemaker) 46
Concentrate (Home Winemaker and Commercial Winemaker) 46
Rectified Grape Juice Concentrate 47
Juice Storage 48

Chapter 2: Grape Processing 49

Equipment 49
Stemming and Crushing 56
Juice Separation 58
Skin Contact 64
pH Regulation 66
Settling Contact Time 67
Use of SO_2 67
Juice Separation (Home Winemaker) 71
Drainers and Presses 74
Oxidation 79
Pectic Enzymes 79
Juice Additions and Corrections 83
Noble Rot 84
Grape Concentrate (Home Winemaker) 84
Grape Juice Composition 85

Chapter 3: Fermentation and Wine Composition 92

Yeast 92
Inhibition 99

Yeast Growth in and Fermentation of Juice 103
Fermentation Biochemistry 127
Wine Composition 137

Chapter 4: Clarification and Fining of Wines **146**

White Wines – Fining 146
Red Wines – Fining 150
Clarification and Filtration 154
Home Winemaking Filtration 166

Chapter 5: Stabilization **168**

Oxidation 168
Inorganic and Organic Precipitates 171
Microbiological 183

Chapter 6: Secondary Fermentations **186**

Malolactic 186
Flor Sherry 194
Champagne or Sparkling Wine 201

Chapter 7: Aging, Bottling and Storage **215**

Tainted Wines 215
Storage and Aging 216
Bottling 231
Storage and Transport 242
Home Winemaking Bottling 243

Chapter 8: Sensory Evaluation **244**

Evaluation of the Wine 244
Scoring Systems 249
Descriptive Terminology and Analysis 252
Statistics 255
Winery Tasting 256

Chapter 9: Chemical Analysis and Information Retrieval **258**

Analysis 258
Grape Juice 258

Wine 264
Water Tests 270
Information Retrieval 271

Chapter 10: Additives and Contaminants **273**

Additives in General 273
Legal Additives 275
Illegal or Unused Yeast Inhibitors 281
Compounds Added for Flavor or Compositional Change 285
Natural Inhibitors 287
Special Natural Wines and Coolers 288
Fraud Detection 288

Chapter 11: The Home Winery **291**

Crushing and Fermenting Building 291
Facilities 292
Equipment 292
Cellar 295
Laboratory 299

References **300**

Bibliography **318**

Appendixes **321**

Index **327**

Preface

This book is for the winemaker in a medium or small operation who plans to make table wines or champagne. It is also for the intelligent home winemaker who wishes to know more about winemaking than adding two teaspoons of this or that and stirring with a clean stick. Without some background in chemistry and biochemistry, it may be difficult in places. An effort has been made to explain in the simplest terms possible and still not forsake the real meat of the problems.

A very abbreviated description of planting and growing a vineyard is offered. This is aimed primarily at the home wine enthusiast. What is said applies to all viticulture, however.

Some of the chapters are more intense than others. To a degree this reflects the author's personal interests. An effort has been made to be complete and up-to-date in each area.

The Bibliography covers most of the current texts that should be of interest to the winemaker. If some favorites of others are omitted, it is because what was given seems to cover the field adequately. References older than 1980 were kept to a minimum. Emphasis has been on the literature published after that date. Most of the literature to that time has adequately been covered in the texts cited in the Bibliography.

The analysis chapter is complete for routine analyses. Other texts are available for the more complex analyses. Commercial winery design was not covered because of the complexity and the need for specific needs for specific wants. Waste removal, by-products such as yeast, color, tartrate recovery, distillation, brandy production, dessert wine and specialty wines are not covered. Government regulations are readily available by request from the proper agency and are always in flux and need to be followed consistently.

This book is meant to be a guide, not a bible, to table winemaking. The differences in style that can result from the variations that can be applied in the winemaking process are for the winemaker to choose.

Cornelius S. Ough

Introduction

Wine types have changed rather rapidly as have the technologies associated with the various wine types. This has led to a shift in production of dessert wines to table wines around the world. In the United States, this has been a very dramatic change in wine consumption (see Table A). The reasons for this are perhaps because of technology more than anything else. Before the 1950s, good drinkable dry table wines could only be made from grapes grown in the limited, cooler, coastal regions. With improvements that did not alter the sensory aspects of the wine, such as cool-controlled fermentations, cultured yeast inocula, sterile filtration and cold sterilization chemicals, good table wines were made from the warmer areas. The wines from the colder areas improved in quality as well. For those who were familiar with the "vin ordinaire" of Europe before the 1960s or 1970s, it was obvious that significant changes took place there also.

Table A. Change in Wine Consumption Pattern in the United States

	Gals/capita		
Type	1950	1979	1987
Table wines	1.08	2.56	2.01
Dessert Wines[1]	0.77	0.44	0.34
Other wines[2]	--	--	1.09

[1]All wines over 14% v/v alcohol.
[2]Includes coolers, natural special wines under 14% alcohol, sparkling wines.

In California, growers changed the grape varieties that they planted to better suit the need for the wine market. These changes reflected the demand for the varieties that produced the highest quality table wines. This was not unique to California, but California was the leader in this area. The newer grape-growing areas, such as Australia, South Africa, and some older ones, have followed suit. A great deal of credit for these changes has to be given to work done by Amerine and Winkler (1963) in their extensive evaluation of the growing regions and their variety recommendations. Table B shows how the variety plantings have changed since their influence was felt. Since this survey, the trend toward planting prestigious table wine varieties continues. Others certainly contributed also. Leadership of industry people like the Gallos had a deciding impact. They took the recommendations to heart and also did their own testings, and insisted on planting the better wine varieties.

These changes have not been as dramatic for the older established areas of Europe. The variety selection has gone on for hundreds of years and the optimum grape cultivars were known and planted. However, the impact of change has taken its toll. The dessert wine and sherry countries have a much smaller share of the world wine market. Extensive plant modification and upgrading has taken place in Italy and France. The Balkan countries are beginning to modernize their plants and make some changes in what is planted. As with everything, these changes come with a price – that price is evident when the consumer goes to the store. The increased prices and better production practices of growing grapes, plus changing social habits, have caused a worldwide glut of wine.

The winemaker in a competitive winery must still strive to improve and make a wine that is unique yet has high quality with certain recognizable features. How well he/she succeeds in this determines only partially the probability of success. How efficiently the plant is operated and how expertly the product is sold to wine writers, distributors, and the public, are also factors in the winery's economic survival.

Many times the winemaker is the only scientifically trained person in the winery. The training must include a broad knowledge of chemistry, biochemistry, bacteriology, a working knowledge of statistics, engineering, business and management skills. Also, one

Table B. Acreage Changes in Wine Cultivars in California Over 42 Years[a]

Cultivar	1944	1986
Whites		
Burger	2987	2496
Chardonnay	<500[b]	29319
Chenin blanc	<500	40088
Colombard	1480	NA[c]
Emerald Riesling	0	2583
French Colombard	<500	69713
Gewurztraminer	<500	3659
Gray Riesling	<500	1761
Malvasia Bianca	NA	2237
Muscat blanc	NA	1531
Palomino	5072	2372
Pinot blanc	<500	2031
Saint Emilion	<500	1009
Sauvignon blanc	<500	15193
Semillon	<500	2724
Sylander	574	725
White Riesling	<500	8449
Other whites	8780	274
Total	23893	186164
Reds		
Alicante Bouschet	25606	2858
Barbera	<500	13417
Cabernet Sauvignon	<500	23149
Carignan	32051	14347
Carnelian	0	591
Gamay	NA	2282
Grenache	4229	14669
Mataro	7692	456
Merlot	<500	2881
Mission	10906	1883
Petite Sirah	7721	4983
Pinot Noir	<500	10842
Royalty	0	999
Ruby Red	0	7334
Ruby Cabernet	0	9319
Salvador	NA	922
Tinta Madeira	<500	739
Zinfandel	50349	25797
Other reds	16747	468
Total	157801	137936

[a]1 acre=0.4047 ha.
[b]<500 was taken as 330 for the purposes of total acreage.
[c]NA=Either less than 500 acres or listed under 'Other whites' or 'Other reds'.

should have a thorough understanding of viticulture including knowledge of cultivar identification, maturity testing of grapes and fruit quality. Consultants are available but expensive, so one must decide based on his/her experience and training. This can be a very lonesome job when a 25,000-gallon tank of Chardonnay develops a peculiar odor or taste. One must make a plan to abort a potential problem and cure the ill. Is it due to a yeast or bacteria or is it some chemical action? Did one of the workers drop a wrench into the tank, etc.? The tank of wine could be worth $100,000 or more to the winery but, if spoiled, only a fraction of that. Nowadays it is a very scientific and serious business.

Chapter 1

Grapes and Concentrate

The source of the raw material for winemaking is of vital importance to the quality of the wine produced. Several options exist for winemakers or winery owners: (1) grow their own grapes, (2) buy grapes from local vineyards, or (3) buy grapes to be shipped to them. Of all the choices, the first can be the most rewarding if local growing conditions are favorable. The winemaker needs a moderate understanding of grape-growing (viticultural) practices. Most commercial wineries deal only with options (1) or (2). They generally grow their own grapes, have a trained viticulturist or hire an appropriate vineyard consultant to advise them.

VINEYARD

Before planting even a small vineyard, there are many variables to consider. The selection of the scion (top) of the vine and the rootstock onto which it will be grafted must be such that the two are compatible with each other. They must be compatible with the climate and the soil. The following questions must also be answered. Do pests, such as phylloxera, Pierce's disease, or nematodes exist in the area? Is the rootstock resistant to the soil pests? Is there adequate soil depth, drainage or moisture in the soil? Is there freedom from chemical imbalance? Will the location provide adequate sunlight? How does one find the answers to these and other questions?

The best solution is to seek qualified advice. In areas where grapes are grown, the local farm advisor or county agricultural agent will be able to answer the questions, generally at no cost to the winemaker. In an area where grapes are not grown, then it would be wise to seek the help of a qualified viticulturist. It would also be prudent to obtain a textbook on grape growing such as *General Viticulture* [Winkler et al. (1974), University of California Press, Berkeley, California].

From the time a vineyard is planted until the first good crop of grapes is harvested is from four to five years. Under favorable growing conditions it is possible to obtain a crop in three years. The fruit from the young vines is often chemically imbalanced and the wine is not a true representation of the vines' potential. Overcropping young vines can do serious harm to the vines, even to the point of stunting or killing them.

There are certain clichés which are worth considering: (1) if grapes are not grown in your area, there is probably a very good reason for it, and (2) the wine you make will only be as good as the grapes (or material) it is made from.

Preparing for and growing a vineyard can be a costly project. It can also be a very rewarding one both financially and personally. The personal aspects include "your wine from your grapes." Harvest time is usually August to September or October. The leaves are turning to bright colors, the weather is mild and pleasant with a feeling of fall in the air. Memorable times can result.

Most high quality grapevines planted in wine-growing areas yield about three to seven tons of grapes per acre depending on the variety and the location. Most home winemakers can expect to get 80-110 gallons of juice per ton of grapes using the equipment available to them. With the more sophisticated presses, yields up to 140-160 gal/ton are possible. Yields on red grapes fermented on the skins will be about 20% higher because of the ease of pressing. Planting more than one acre of grapes will put the home winemaker over any reasonable amount of wine for home use. Depending on the spacing between the vines, an acre will contain 500-600 grapevines.

Growing Conditions

There are several growing conditions which will affect the composition of the grape and the quality of the wine. Among these factors are climate, soil, fertilization, irrigation, pest control, virus infections, crop level and other cultural practices.

Climate is probably the most important single identifiable factor which can alter the final wine quality. Grapes grown in warm climates tend to lose their acidity more rapidly than grapes grown in cooler regions. This is due primarily to the enzymatic decomposition of malic acid during the maturation of the grape. The warmer the climate, the more rapidly the loss of malic acid occurs. This is one of the many biochemical changes taking place in the grape as it ripens. The tartaric acid, the other major acid in grapes, is more resistant to decomposition and less change occurs. Most of these changes are caused by enzymes and each has an optimum temperature at which the most intense enzymatic activity takes place. Another example of the effect of climate is how the varietal character of Sauvignon blanc is altered by climate. In the cool vineyards of Monterey County in California, the Sauvignon blanc has a strong and distinct character when it reaches full sugar maturity. On the other hand, in the warm San Joaquin Valley in California you can find the same varietal character, at lesser intensity, when the grape is about two-thirds mature; but the varietal characteristic has disappeared when full sugar maturity is reached. In some cases, the varietal characteristic is so intense in the cooler climates that it precludes growing it in those areas.

Grapevines need to go dormant to survive the winter cold. This is best accomplished by a cool fall with light to medium frosts which allows the vine to store sufficient carbohydrates to sustain adequate spring growth. If these conditions do not exist, then the vines are easily damaged by sudden intense winter cold and may be killed. On the contrary, in the warm semi-tropical areas, they may go only partially dormant. Insufficient carbohydrate reserve will be stored for adequate growth and fruit set the following year.

Climatic conditions affect the degree to which the grape or grapevine will be subject to mold, rot or mildew. For example, White Riesling will rot on the vine before it is mature in the warm and heavily irrigated vineyard areas. In any area that has summer rains or is heavily irrigated, the vines are subject to powdery mildew (oidium) or worse, downy mildew (peronospora) under the wettest conditions. Fruit quality is badly impaired when molds are not controlled. When overirrigated, varieties with tight clusters such as Zinfandel have a tendency for the berries to swell and burst. This allows mold and rot to occur. Usually the berries that rupture are in the center of the cluster and the entire cluster becomes contaminated. This can be avoided by planting these tight cluster grapes in the dryer areas and avoiding irrigation as the berries reach full size.

Soil Pests

The soil pests most common in the northern hemisphere are nematodes and phylloxera. Nematodes (round worms), usually found in sandy soil, suck nutrients from the roots of the vines. Poor growth and delayed grape maturity result. Such stunted vines seldom yield good quality grapes. Nematodes are controlled by fumigating the soil with chemicals before planting the vines. Choosing nematode-resistant rootstock is necessary. Certain nematodes also spread viruses which affect the grapevine. Phylloxera is a root louse (insect) native to the eastern United States. The native American grapes were shipped to various areas of the world for experimentation. These vines were not sterilized before shipment so they carried the phylloxera with them. This caused great damage in Europe and other grape-growing areas in the middle to late 1800s. It was a great economic tragedy. The life cycle of the pest is quite interesting and complex. In the grape-growing areas of the world that receive little or no summer rains (California, Israel, South Africa, etc.), the winged form of the pest is not present. The infestation of this pest then can be noted by an ever-widening circle of dead or stunted vines from the point of infestation. These pests also suck nutrients from the roots and are more prevalent in the heavier soils than in the light sandy loam-type soils. Again, the treatment is soil fumigation and proper selection of rootstocks. A graft of a resistant rootstock to a scion is shown in Figure 1-1. Data from a typical rootstock trial is given in Table 1-1. This was on Ballard clay, cane pruned, with set sprinkler irrigation (Foott et al., 1989). Under other conditions, different results may be obtained.

Soil

In Europe, soil is considered a major factor in determining wine quality. Scientifically sound proof of this remains to be shown. Conditions such as extreme acidity or alkalinity will cause problems

FIGURE 1-1. A vine showing the graft joint a few inches above the soil. (If graft is below the soil, roots from the scion may form.)

Table 1-1. A Comparison of Rootstock Interactions on Pruning Weights and Crop Levels for Cabernet Sauvignon and Chardonnay Scions.

	Pruning Weights (lb/vine)		Crop Yield (lb/vine)	
Rootstock	Cabernet Sauvignon[1]	Chardonnay[2]	Cabernet Sauvignon	Chardonnay
1202C	6.5[a]	2.7[a]	14.1[ab]	20.6[ab]
5A	4.5[b]	2.1[b]	13.5[ab]	21.6[a]
St. George	4.4[bc]	1.3[c]	12.4[b]	16.0[bc]
SO_4	4.2[bc]	1.5[c]	14.4[ab]	15.5[c]
Own	4.1[bc]	1.8[bc]	13.5[ab]	17.9[abc]
3309C	4.0[bc]	1.6[c]	12.6[ab]	19.0[abc]
AxR#1	3.9[bc]	1.8[bc]	15.8[a]	21.7[a]
110R	3.2[bc]	1.3[c]	12.5[b]	15.1[c]
Harmony	3.0[c]	1.3[c]	9.1[c]	9.5[d]

[1]Average of four years' harvests.

[2]Average of three years' harvests.

[a-d]Values with the same superscript in the vertical columns are not significantly different.

From Foot et al. (1989)

in vine growth. This, in turn, affects the wine quality. Normally, it is adequate if the soil is three or more feet deep, has good drainage and is at a reasonable pH. Other deficiencies can usually be remedied by proper treatments and/or additions. It is best to have an expert evaluate the soil if there are any doubts about planting grapes. At that point the need for, and type of, irrigation should be considered. Additions of fertilizer should be moderate. Most soils have adequate nutrients. For soils with depleted nitrogen, additions of over 50 pounds of nitrogen per acre are wasteful as excess top growth can result with no gain in crop. It can also lead to excess nitrogen compounds in the wine which can encourage spoilage as well as possible formation of ethyl carbamate.

Planting

Verification is made that the conditions of the climate and soil are adequate for a vineyard. The soil is plowed, disced and fumigated (if required). The field is laid out in rows using a surveyor's transom. The easiest type of plants to use is grafted plants in pots. These can be planted anytime in the spring after danger of frost has passed.

In shallow soil it is useful to lay black plastic around the vines to hold in moisture and heat. It will also help control weeds. If water is available, several good irrigations will help establish the root system sooner.

The usual insect and fungus protection should be applied to the young vines.

It is important that reference to viticultural literature be made or consultation obtained so that a thorough awareness of all facts of vine planting and maintenance is available before planting.

Irrigation

Grapes are deep-rooted plants and require little irrigation after the vines are established. Over-watering will encourage excess cane growth, delay in fruit maturation and poor carbohydrate storage for winter and spring vine reserves. If the soil is shallow, more frequent but smaller volume application of water should be made. Mature vines planted in deeper soil require only one or two heavy irrigations during the summer in the warm areas and none in cooler areas. If the grapes are grown where summer rains occur, no additional water applications should be made. If too much water is used, the delay in sugar maturity allows acidity and flavor losses to occur. Also, picking may be delayed until the rainy season which allows the opportunity for mold and rot to infect the grapes.

It should be emphasized that outside consultation is recommended for the amateur grape grower if any questions arise.

Pruning, Trellis and Yield

The proper time to prune your grapevines is in the winter after they have gone dormant and the canes have hardened (turned brown). There are several styles of pruning: head (or bush), cordon and cane. General schematic examples of each of these styles are shown in Figure 1-2. Although other pruning and trellis styles may give better yields, for the home vineyard the simplest is the most satisfactory.

Each vine should be staked and the vine trained up the stake during the first growing season. Only one shoot should be trained and all the others pinched off. Figure 1-3 shows a one-year-old grafted vine ready to train. Figure 1-4 shows a typical three-year-old grafted vine that has been trained up the stake and onto a two-wire horizontal trellis. An example of a vertical two-wire trellis for cordon-spur pruned wines is shown (Figure 1-5).

The number of buds, canes or spurs to leave on the vine depends on the size and vigor of the vine. More buds may be left on the larger and more vigorous vines. If an extremely large crop sets, then some thinning of the clusters may be necessary. Actual thinning should be considered carefully. Ough and Nagaoka (1984) found that when a normal crop could be matured to a desired °Brix and pH, thinning did not increase wine quality. The loss in quantity

FIGURE 1-2. Schematic drawing of A) cane pruned vine, B) head pruned vine, and C) cordon-spur pruned vine.

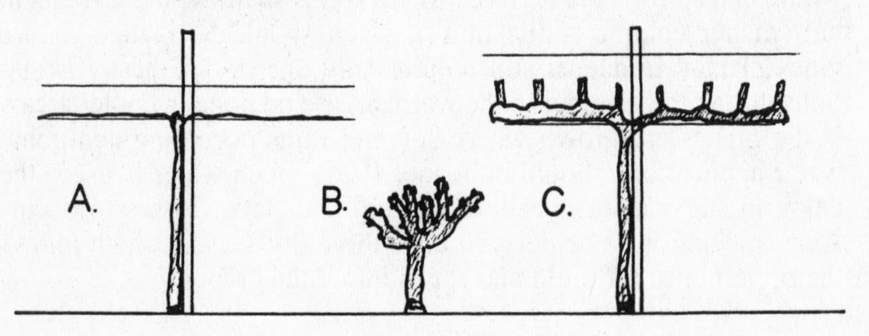

FIGURE 1-3. A young grafted vine ready to cut back to one shoot and train up the stake.

FIGURE 1-4. A three-year-old vine trained onto a two wire horizontal trellis.

FIGURE 1-5. An example of cordon-spur pruned vines on a two-wire vertical trellis with drip irrigation.

of fruit served no purpose. Differences in vineyards, even adjacent ones, cause bigger quality differences in the wines. In the coastal areas with shallower soils, one can expect to get roughly about half to two-thirds the yield of the richer valley soils. Table 1-2 gives some general viticultural data gathered for a University of California Bulletin. Data are for conservative vineyard management to obtain optimum fruit quality in the coastal areas.

GRAPE MATURITY

Most people think of fruit maturity in terms of the sugar content of the grape. This is true in the warmer areas where grapes will accumulate as much as 26 grams of sugar per 100 ml of juice (°Brix value of 28°). Care is usually taken to pick the fruit for white table wine at 20.5°-22.5°Brix and the red table wine grapes at 21.5°-23.5°Brix. There is a very practical reason for this. If harvested in these ranges, the wine will have sufficient alcohol to give a sound wine which is balanced in taste and resistant to microbiological

Table 1-2. Yield Data for Some Common Wine Varieties Grown in Central Valley and Coastal Areas of California.

	Tons/acre			Tons/acre	
Cultivar	Central Valley	Coastal Area	Cultivar	Central Valley	Coastal Area
Barbera	8.2	4.3	Grenache	8.3	6.2
Cabernet Sauvignon	7.0	4.5	Pinot Noir	3.7	3.9
Carignane	9.0	-	Ruby Cabernet	7.9	5.1
Chardonnay	6.3	4.4	Sauvignon blanc	5.8	6.4
Chenin blanc	10.0	5.8	Semillon	6.4	3.9
French Colombard	7.1	5.9	White Riesling	4.9	4.3
Gewurztraminer	5.4	6.1	Zinfandel	-	4.1

Data from Amerine and Winkler (1963). Coastal area fruit was non-irrigated and on two-wire trellises.

(If irrigated and larger trellises, increased crops would be expected.)

spoilage. Table 1-3 gives the approximate yields of ethanol that can be expected from white and red grapes picked at various °Brix values from cool and warm grape-growing areas.

Brix measurement, °Brix, can be measured with either a hydrometer or a refractometer. The refractometer is the most commonly used device for measuring sugar in grapes and the better models are temperature compensated. Less expensive models are not temperature compensated and the readings have to be corrected for temperatures different than those for which the refractometer is calibrated. The method for temperature corrections is given in Appendix III. Even with the temperature-corrected model, it is best to work as closely to the calibrated temperature as possible. You should also check your refractometer at the beginning of each season for accuracy.

Method for Checking Correctness of Refractometer

Weigh out 20 grams of sugar and dissolve in 80 grams of water. (If a gram scale is unavailable, measure out four ounces of sugar

Table 1-3. Effect of °Brix and climate on alcohol yield.

	Ethyl alcohol % v/v		
		Red grapes	
Brix	White Juice	Cool area	Warm area
18	10.0	9.7	9.0
19	10.7	10.3	9.6
20	11.3	10.8	10.2
21	12.0	11.4	10.8
22	12.6	12.0	11.2
23	13.2	12.7	11.9
24	13.9	13.3	12.4
25	14.5	13.8	13.0

and dissolve it in one pound of water.) This will give a 20°Brix solution. The refractometer should read 0° with plain water and 20°Brix with the sugar/water solution. If it does not give these readings, the instrument must be recalibrated. If the instructions do not give a method for calibration, return the refractometer to the factory for calibration.

Proper Care of Refractometers

Care should be taken to avoid scratching the prism (where the juice sample is applied). Never use a glass rod or other hard applicator. Use a soft applicator such as a rubber-tipped plastic rod. The prism should be dry when the sample is applied. Errors occur in refractometer readings when moisture is left on the prism, diluting the sample. If a previous sample is left to completely or partially evaporate, the residual sugar can cause a new sample to give a falsely high reading. The prism should be rinsed thoroughly and only lens tissue used to dry it.

Vineyard Sampling

Vineyard sampling should be done carefully and correctly. The following method is used by professional viticulturists. It is a scientific method and is simple to follow. Home winemakers and/or grape growers can easily make this evaluation if they are willing to spend the necessary time. For long-range planning, a record could be kept of these readings. Mark those vines chosen for sampling so that a new vine selection each time is unnecessary.

1. Select portions of the vineyard that will represent the entire vineyard. Variables such as soil depth (deep or shallow), vine age (young or old), etc., should be considered.
2. Choose percentages of vines from each which represents each group fairly.
3. Sample the designated or random vines by selecting 10 to 20 clusters or enough to have a representative selection from the vineyard. The clusters should include some near the trunk, middle and outer portions of the vines as well as the shaded and sunny areas.

4. Pool the cluster samples. Crush and press them in a manner similar in severity of pressing to the method or processing that will occur when they are harvested. Be sure the resulting juice is well mixed.
5. Determine the °Brix. This is based on the recommendations of Kasimatis and Vilas (1985). Previous recommendations suggested use of berry sampling and other modifications which took extra time and gave little or no more accuracy. Figures 1-6 and 1-7 show the type of error to result from rational sampling. Sampling should begin at 16°Brix and weekly samples taken. It is worthwhile to plot the changes in °Brix against the intervals in time between samplings. A typical plot of these changes is given in Figure 1-8. This will allow one to predict the date of harvest several weeks in advance if data from previous years are available. It is also valuable to plot the total acidity and pH during the maturation period as shown. After a few years of data collection, the optimum harvest conditions can be determined. This is not absolutely necessary, but if the winemaker can make the latter two determinations, it will allow him/her the options of adjusting the acidity or pH.

A method which has been reported (Cooke and Berg, 1983) as a more popular method than the °Brix/acid ratio and the suggested replacement method of assessing the proper harvest time, is the °Brix × $(pH)^2$. This more properly relates to the true physiological maturity of the fruit. Whatever the criteria, the fruit for table wine should be harvested in a practical °Brix range. White table wine grapes are ready to harvest around 22°Brix and red table wine grapes about 23°Brix with pH values of 3.2-3.4 and 3.3-3.5, respectively. Alterations of the total acidity and pH can be adjusted by acid additions. In some states or countries, the °Brix can be altered by sugar additions but not in California.

Analysis

In this book, we give only the necessary simple analytical procedures. These can be done with a minimum knowledge of chemistry and require minimum equipment. These procedures are either given in Chapter 9 or in appropriate places throughout the text.

FIGURE 1-6. °Brix variances calculated for a range of berry sample sizes. Dotted lines indicate variance spread from the mean (Kasimatis and Vilas, 1985).

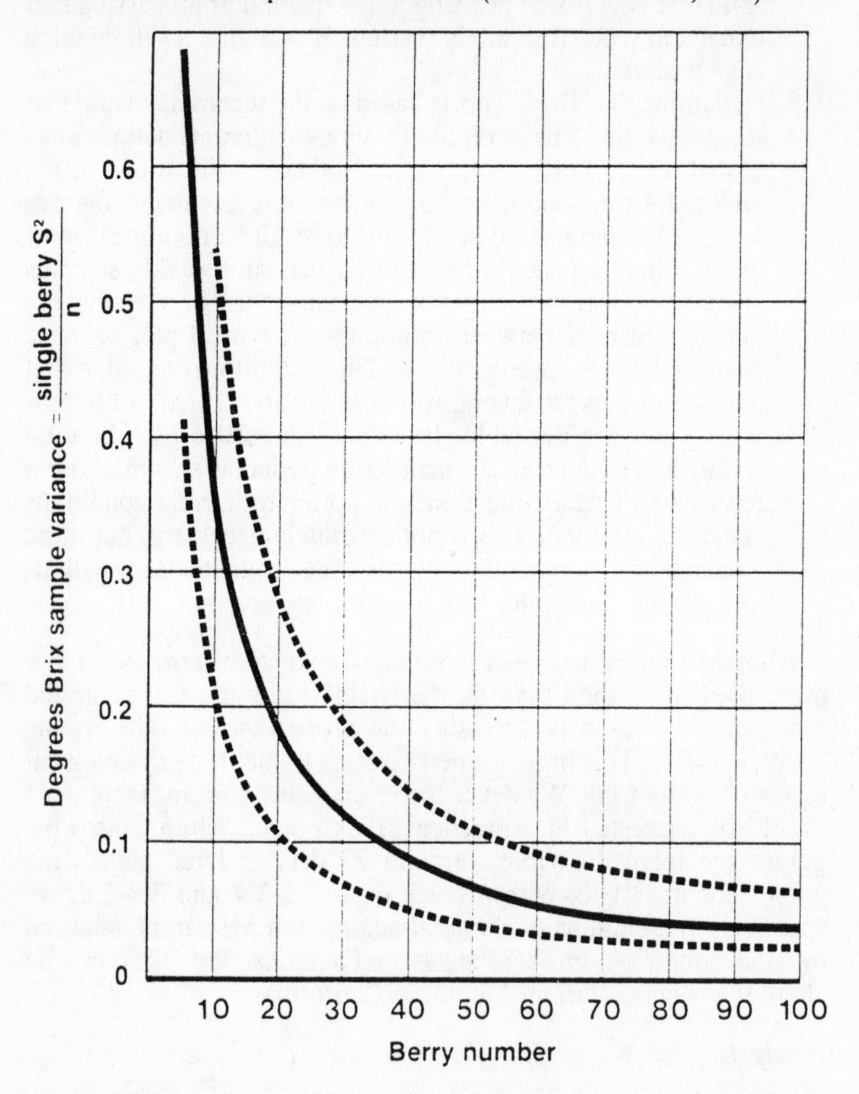

Degrees Brix variances calculated for a range of berry sample sizes. The solid line is the mean and the space between the broken lines is the expected variance spread.

FIGURE 1-7. °Brix differences calculated for various sample and replication numbers at 80% probability for detecting the differences at the 5% level of significance (Kasimatis and Vilas, 1985).

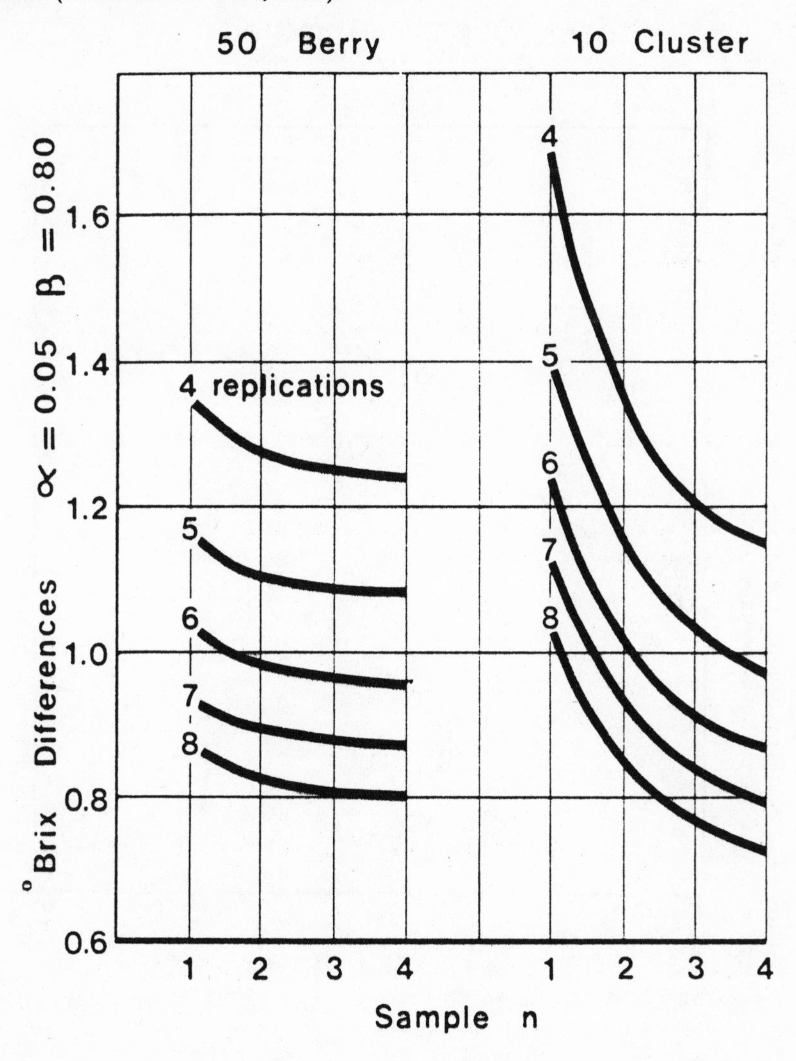

Degree Brix differences calculated for various sample and replication numbers at 80% probability for detecting the differences at the 5% level of significance.

FIGURE 1-8. Plot of °Brix and total acidity of the maturing grapes with time. From this the expected date of harvest can be partially anticipated.

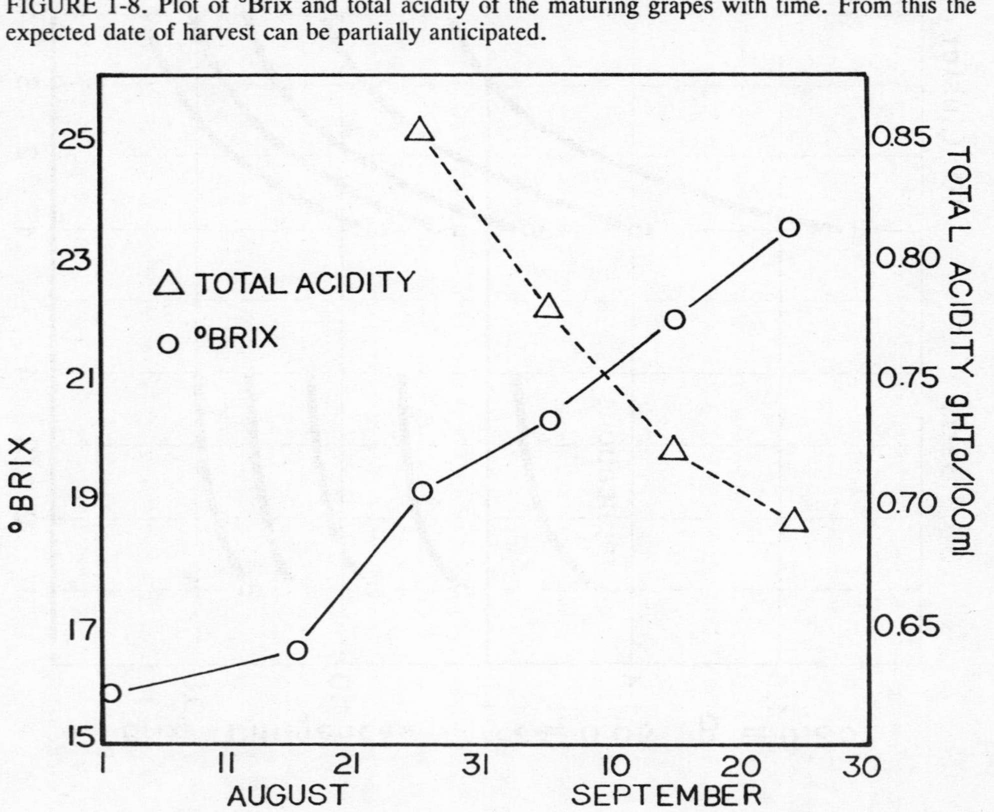

WINE GRAPE VARIETIES

It might be informative to consider what is meant by genus, species and variety (cultivar). *Vitis* is the major classification or genus. The main grape species used for wine are *vinifera* (the one used to the greatest extent for wine throughout the world), *lubruscana* (such as Concord) and *rotundifolia* (such as the muscadine). The latter two have limited use for wine. These and other *Vitis* species (not used directly for wine) have been hybridized (crossed) with *vinifera* to give what are commonly known as "French hybrids." Most of the rootstocks used are pure or hybridized *Vitis* species other than *vinifera*. Variety or cultivar is the final designation. The scientific name of Cabernet Sauvignon would be *Vitis vinifera* var. Cabernet Sauvignon. This simply means it is a variety within the species.

There are hundreds of grape varieties. If hybrid varieties are included, there are thousands. Only those varieties that are considered commercially acceptable for table wines will be emphasized, but others that interest home winemakers will also be discussed.

The desirable qualities in a grape variety are:

1. The grapevine must be able to grow and mature a reasonable crop of fruit.
2. The fruit must tolerate the climate and soil conditions.
3. The fruit must maintain a reasonable balance of acid and sugar at maturity.
4. The fruit should have a recognizable and desirable character when harvested and made into wine.
5. Drastic color changes after vinification must not occur.

If the above conditions are met, then the number of varieties one should grow is limited. Table 1-4 gives some viticultural characteristics for some commonly used cultivars.

Climate variations are reasonably easy to classify in California. Most areas have no summer rain, cool nights and either hot or warm days, and a low potential of frost damage. In the eastern United States, because of high amounts of summer rainfall and extremely cold winters, there are limitations on vine growth and fruit productions of *vinifera* varieties. It is extremely difficult to grow *vinifera*

Table 1-4. Viticulture Characteristics of Wine Grape Cultivars.

Variety	Production (tons per acre) (for central valley)	Harvest period	Vigor	Harvestability		Some Important Characteristics
				Hand	Machine (juicing)	
White table wine varieties						
Burger	9-12 (as high as 16)	Late	Moderate	Easy	Hard (very heavy)*	Susceptible to bunch rot and overcropping. Total soluble solids tend to be low even with light crops.
Chenin blanc	9-12	Middle	High	Medium	Medium (medium)	Tight clusters can contribute to bunch rot. Minimized by early irrigation cut-off; prebloom gibberellin application, and use of fungicides. Best in fine sandy loams.
Emerald Riesling	9-13	Middle	Exceptional	Medium to hard	Very hard (very heavy)	Crushed grapes very susceptible to browning, requiring special care during harvest. Adaptable to all soil types.
French Colombard	9-14	Late	Exceptional	Medium	Medium (medium)	Excess vigor can be a problem on deep, fertile soils. Dense foliage can interfere with sulfur dusting and harvest operations. Occasional bunch rot associated with powdery mildew problems or Botrytis infections. May need Zn correction.
Gewurztraminer	6-8	Early	Moderate	Medium	(medium)	Small clusters.
Green Hungarian	12-15	Late	High		(heavy)	Susceptible to bunch rot. Irrigation cut off prior to harvest reduces bunch rot and breakdown. Fails to ripen properly.
Grey Riesling	8-10	Early	Moderate	Medium	(medium)	Ripens rapidly at harvest time.
Peverella	8-11	Late	High	Easy to medium	Unknown	Compact clusters sometimes subject to rot. Dense foliage.

Saint-Emilion	10-13		High	Easy	Medium (heavy)	Fruit "holds" well on vine.
Sauvignon blanc	5-7	Early-Middle	Exceptional	Hard (small clusters)	Medium (medium)	Excess, dense growth can interfere with sulfur dusting and harvest operations. Susceptible to Botrytis rot from fall rains.
Sauvignon vert	8-10	Middle	High	Medium		Fails to properly matue to desirable sugar. Acid is low.
Sémillon	7-11		Moderate	Medium	Medium to hard (very heavy)	Rot is sometimes a problem if the fruit is held on the vines too long.
White Riesling	3-5	Early-middle	Weak	Medium	Medium (medium)	Fruit breaks down rapidly when held beyond 19°B. Susceptible to bunch rot.
Red table wine varieties						
Alicante Bouschet	8-10	Late	Moderate	Medium	No information	Compact clusters sometimes subject to bunch rot. Vines are sensitive to overcropping.
Barbera	6-10	Late	Moderate	Easy	Medium (medium)	Leafroll virus is widespread in vineyards originating from non-certified planting stock. Vineyard establishment from cuttings often results in below normal initial stands.
Cabernet Sauvignon	5-7	Middle	Moderate	Medium	Easy to medium (light to medium)	Few problems but a very moderate producer. Tends to lack good fruit color in warm areas.
Calzin	5-11	Middle-Late	Moderate	Medium	Unknown	Resistant to red spider. Loose clusters and little rot.
Carignane	8-13	Middle-Late	High	Medium	Medium to hard (hard)	Susceptible to powdery mildew. Summer bunch rot problems in some vineyards require pre-bloom gibberellin treatment. Yellow vein virus "unfruitful" vines found in many commercial plantings.
Gamay (Napa)	5-8	Middle	Moderate	Medium	Medium (medium to heavy)	Larger, compact clusters are subject to bunch rot.
Gamay Beaujolais	See Pinot noir					

TABLE 1-4 (continued)

Variety	Production (tons per acre) (for central valley)	Harvest period	Vigor	Harvestability		Some Important Characteristics
				Hand	Machine (juicing)	
Grenache	8-14	Middle	Very high	Medium	Medium to hard (medium)	Bloomtime Botrytis infections are sometimes a problem. Compact clusters sometimes rot. Subject to delayed bud break in spring and occasional poor fruit set. Good tolerance to salt and boron.
Grignolino	No recent experience					
Malbec	5-9	Late	Moderate	Medium	Medium (medium)	Subject to poor fruit set under high nitrogen conditions.
Niabell	3-5	Early	Low to moderate	Medium	Easy (light)	Berries subject to uneven ripening. Exposed fruit sunburns during hot spells. Fruit shrivels if harvest is delayed. Resistant to powdery mildew.
Pinot noir	4-5	Early	Low to moderate	Medium to hard	Medium (medium)	Tight clusters subject to bunch rot, especially with delayed harvest. Fruit raisins badly with delayed harvest.
Pinot St. George	No experience					
Petite Sirah	6-10	Late	Moderate	Medium	Medium to hard (medium to heavy)	Tight clusters are subject to rot. Fruit tends to sunburn and raisin from hot spells or delayed harvest. Leafroll and corky bark vines prevalent in older plantings.

Rubired	8-12	Late	High	Medium	Easy (medium)	Tolerant to berry mildew. Young vines subject to collar rot. Young vines easily overcropped. Many commercial vineyards are infected with leafroll virus. Crop estimate is difficult because of high percentage of stems per cluster.
Ruby Cabernet	6-10	Late	Moderate	Hard	Medium to hard (medium to heavy)	Performs very poorly in soils with limiting factors such as sandy texture, compaction, nematode problems, or shallow depth. Irregular berry set occasionally. Many commercial vineyards are infected with leafroll virus.
Salvador	7-11	Middle-Early	Moderate	Hard	Easy (medium)	Resistant to powdery mildew and somewhat resistant to hoppers, mites, and grape leaffolder. High thrips populations sometimes stunt shoot growh. Tight clusters subject to bunch rot. Harvest should not be delayed. Difficult to prune--bushy growth.
St. Macaire	7-10	Late	Moderate	Easy to medium	Medium to hard (medium to heavy)	Grapes tend to sunburn if vines are dried up early. Medium-to-large clusters.
Valdepeñas	10-12	Middle	High	Easy to medium	Medium to hard (medium to heavy)	Heavy, dense leaf area and powdery mildew susceptibility requires especially good sulfur dusting practices. Vines leaf out late.
Zinfandel	5-9	Middle-Late	Low to moderate	Medium	Hard (medium to heavy)	Susceptible to spider mites. Bunch rot problem during ripening requires pre-bloom gibberellin spray. Exposed ripening fruit subject to raisining.

in the northern and plain states except in those areas where the climate is modified by the Great Lakes and in Washington, Oregon and Idaho where the winter storm patterns usually bring reasonably warm air. In the southern states, the summer rains increase the danger of mildew, mold and rot. The climate allows the host of Pierce's disease to over-winter and intensifies this problem. Certain grape varieties seem to be more resistant to winter cold than others. In an area with severe winters, the choice of *vinifera* varieties may be limited to ones similar to those grown in Germany. Recent success with Chardonnay in cold bench areas south of Lake Erie is promising. Wise and Pool (1989) have great praise for the success of Cabernet varieties and other *vinifera* grown in the Long Island area of New York. Even these varieties will not survive the cold winters in some areas of the United States. The only choice left is either French hybrids or the native American varieties.

Labrusca and French Hybrids

There are several *labrusca* and French hybrids that are used for wine. These varieties are often grown in areas where *vinifera* are difficult or impossible to grow. A text on the vineyard management of these varieties should be consulted before planting. A limited list of varieties used for wine in the Ontario areas of Canada and the eastern United States is given below in Table 1-5.

Harvest time will vary from August to the middle of October depending on the area and the crop. To make wine, one must add

Table 1-5. Some Common American Grapes and Hybrids Used in Eastern USA and Canada.

Labrusca	French Hybrids
Catawba	Foch
Delaware	Siebel 9110
Dutchens	Seibel 9549
Elvira	Bertille-Seyne 2862
New York Muscat	Seyne-Villard 172

sugar and water. Plan for a medium crop and allow the grapes to reach their maximum maturity before harvesting. The latter conditions will reduce the acidity and allow for better flavors. Increased tonnage may result in the vineyard because of increased fruit weight.

The varietal character of the *labrusca* grape is from methyl anthranilate and other compounds. The amounts in juice of Concord range from 0.14 to 3.50 mg/L and in wines of these American varieties from trace to 3.1 mg/L. Heating the juice and skins for color extraction (the usual practice) extracts out over half again as much methyl anthranilate. The French hybrids have other varietal-specific components.

Vinifera Varieties

Cabernet-type grapes came originally to the United States from the Bordeaux region of France. The Cabernet Sauvignon has more varietal character and color than other members of that family. When it is grown in a very cool climate, the varietal character takes on a herbaceous aroma that may or may not be displeasing. Bayonove et al. (1975) identified the aroma of Cabernet Sauvignon as 2-methoxy-3-isobutylpyrazine. It is a light-sensitive (Heymann et al., 1986), very potent aroma compound that does not degrade quickly. This may explain why several wineries put their Sauvignon blanc in clear glass bottles. Opening up the vines so the sunlight shines on the clusters lowers the pyrazines. Merlot and Cabernet franc are less intense in color and aroma and pose some viticultural problems. The grapes of these varieties make lighter wine that matures more rapidly than Cabernet Sauvignon, they seldom make the "Great Wine." Malbec is high in malic acid and wines made from this variety sometimes undergo a violent malolactic fermentation. They slowly or never recover from its effects. The choice of varieties for the moderately cool areas is Cabernet Sauvignon and perhaps some Merlot. If the area is hot and a Cabernet type is preferred, the choice is limited to Ruby Cabernet or one of the new *vinifera* hybrids such as Centurian. Both are genetic crosses of Cabernet Sauvignon and Carignane. They produce a good yield of grapes with excellent color that makes a full-bodied and distinctive wine. Carmine, another *vinifera* cross of Ruby Cabernet back-crossed with Cabernet Sauvignon, is showing promise in cooler areas. (See Chapter 2 for acidity additions to must.)

Two white varieties, also from France, are Sémillon and Sauvignon blanc. These two varieties have varietal aromas similar to Cabernet Sauvignon. Sauvignon blanc has, in addition, similar viticultural growth characteristics. Augustyn et al. (1982) first identified the pyrazines of Sauvignon blanc. In the Sauvignon blanc wines from Australia and New Zealand, these varietal components were isolated. They were found to be isobutyl methoxypyrazine, isopropyl methoxypyrazine and secondary butyl methoxypyrazine at about 0.035 μg/L, 0.005 μg/L and 0.001 μg/L, respectively (Allen et al., 1988). The amounts of the first two compounds are well above threshold levels. The method used to determine the pyrazines was described by Harris et al. (1987).

Both Sauvignon blanc and Sémillon are suited for growing in moderate climates but, if grown in too warm a climate, lose their varietal aroma before harvest. If grown in a cold climate, their varietal aroma is almost too intense. The ideal situation is a wine that is both fruity and has a slight but detectable trace of the pyrazine aroma.

Chenin blanc, another French variety, is an excellent variety for the production of a good sound table wine. It has no strong, distinctive varietal character but gives fruity, pleasant wines. This variety bears a good crop and has a wide range of climate adaptability from the warm areas to the cool areas.

Colombard is a variety that can be used for blending because the fruit has a high acid, low pH. It grows best in moderately warm to warm areas. The fruit yield is good and the wine is tart and fruity with little varietal character.

The Pinot noir grape, the main variety of the Burgundy region in France, has not always made outstanding wines from California vineyards. The occasional very good wine produced seldom has the intensity of flavor and aroma associated with the great Burgundy wines. It is best suited to the coolest regions as it ripens rapidly and will raisin in the warmer areas. Interest in Pinot noir clones has resulted in clonal trials. The results of a trial show some differences both in composition of the fruit and wine as well as production.

This trial described is more a trial of Pinot noir selections rather than the true clones (a series of plants propagated from a single vine). Table 1-6 gives the sources; Table 1-7, the average yields, berry weights and cluster numbers/vine by the selections grown in the cool Carneros area of Napa Valley. Tables 1-8 and 1-9 give

Table 1-6. Source of the Wood for Pinot Noir Clonal Selections.

A - Joe Swan - Forestville

B - UCD[1] 3A Wadenswail - 5306-3 10/16

C - UCD 22 Gamay Beaujolais 141 days heat

D - Bealieu Vineyards Block 2, Rows 6 and 7

E - Hanzel Vineyards

F - UCD 1A Wadenswail - Selection B II

G - Chalone Vineyard - Old block

H - Martini Selection 44, Row 17 (High yield)

J - UCD 4 Pommard Selection 820

K - UCD 27 Geisenheim

L - UCD 13 Martini Selection 58 105 days heat

M -Martini Selection 54 (Low yield)

N - UCD 18 Gamay Beaujolais No heat

O - UCD 6 Pommard 119 Days heat

P - Chalone Vineyard New block

R - UCD 12 Pommard Selection 804 89 Days heat

S - Bealieu Vineyards Block 1

T - UCD 23 Wadenswail No heat

V - Martini Selection 58 Moderate yield

Z - UCD 29 (R11, V46, 47)

[1]UCD is University of California, Davis, Plant Foundation Materials.

Table 1-7. Pinot Noir Clonal Selections, Viticultural Data, 6th-9th Leaf Cluster Counts and Weights (1982-1985).

	1982 (6th leaf)				1983 (7th leaf)		1984 (8th leaf)		1985 (9th leaf)	
Selections	Clusters/ vine	Clusters wt (lbs)	Lbs/ vine	Berry wt (g)	Clusters/ vine	Lbs/ vine	Clusters/ vine	Lbs/ vine	Clusters/ vine	Lbs/ vine
A	50	0.226	11.3	0.98	47	10.8	62	17.3	50	11.8
B	64	0.245	15.8	1.05	56	16.4	69	23.3	62	17.7
C	43	0.231	10.0	1.14	45	11.6	48	16.0	48	13.8
D	48	0.234	11.1	1.10	50	11.4	59	16.7	48	12.6
E	42	0.131	5.5	0.74	39	6.0	53	8.5	52	8.2
F	62	0.238	14.8	1.10	65	17.8	61	21.1	67	18.8
G	36	0.167	6.0	1.10	38	6.2	54	10.6	46	6.8
H	46	0.222	10.0	1.10	49	11.3	52	14.9	47	12.0
J	55	0.241	13.3	0.98	47	11.3	56	18.7	46	15.1
K	56	0.295	16.5	1.12	47	12.2	55	18.7	56	19.4
L	54	0.215	11.5	1.04	53	14.5	66	21.7	53	15.0
M	44	0.220	9.75	1.00	47	10.8	62	17.4	52	12.7
N	53	0.245	12.9	1.03	47	12.7	56	18.7	54	16.6
O	44	0.222	9.8	1.00	41	10.6	56	18.0	49	13.3
P	41	0.161	6.6	0.91	38	6.2	52	8.8	45	8.2
R	44	0.236	10.4	0.84	43	9.2	58	16.6	48	12.4
S	35	0.251	8.9	1.07	41	10.5	47	15.4	45	13.0
T	46	0.271	12.6	1.12	5	16.9	59	19.8	61	19.9
V	46	0.217	10.0	1.08	50	12.9	55	16.8	47	12.8
Z	55	0.253	13.8	1.15	47	11.5	58	19.4	51	16.6

some pertinent must and wine analysis. The small-berried selections gave the best quality wines. Several others were consistently above average. Fruit from leaf roll infested vines gave lower quality wines. The size of the berry of red grapes is important and related to wine quality. This is certainly not a new concept but is demonstrated by Champagnol (1984). His data (Table 1-10) show the con-

Table 1-8. Pinot Noir Clonal Selections, Must Analyses. Six-Year Averages and Ranges (1980-1985).

Selection	°Brix		Total Acidity g H_2Ta/L		pH	
	Range	Average	Range	Average	Range	Average
A	22.1-24.5	23.6	8.3- 9.7	8.8	3.13-3.39	3.27
B	21.9-24.4	22.8	7.7- 9.5	9.0	3.10-3.39	3.22
C	21.4-24.4	23.0	7.7- 9.7	8.7	3.13-3.40	3.28
D	21.3-24.8	23.0	8.5-11.5	9.4	3.11-3.31	3.26
E	22.9-25.0	24.6	7.9- 9.5	8.8	3.20-3.41	3.35
F	20.7-23.3	22.1	8.6-10.2	9.4	3.10-3.31	3.22
G	22.8-25.8	24.0	7.7-10.1	8.9	3.10-3.35	3.27
H	20.8-23.7	22.3	8.1- 9.1	8.6	3.12-3.33	3.24
J	22.2-24.3	23.6	7.2- 9.0	8.2	3.15-3.42	3.33
K	21.2-25.0	23.3	7.5- 9.0	8.2	3.12-3.45	3.28
L	20.9-25.3	23.5	7.4- 9.5	8.3	3.11-3.40	3.29
M	22.4-26.4	24.3	8.0- 9.7	8.7	3.15-3.45	3.34
N	21.1-24.4	22.9	7.6- 9.1	8.6	3.10-3.40	3.25
O	22.5-24.7	23.7	6.9- 8.8	7.9	3.20-3.49	3.36
P	22.4-25.7	23.9	7.8-11.3	8.8	3.12-3.45	3.28
R	21.6-25.0	23.6	7.8- 9.0	8.5	3.12-3.50	3.33
S	21.3-24.4	23.0	7.7- 9.7	9.0	3.12-3.40	3.28
T	21.7-25.2	23.1	8.7- 9.9	9.2	3.08-3.29	3.19
V	22.3-24.5	23.4	7.6- 8.5	8.1	3.13-3.40	3.28
Z	22.3-24.8	23.9	7.4- 9.4	8.3	3.12-3.40	3.30

siderable differences in pulp/skin ratios and relative transpiration rates. This shows several positive attributes: (1) more color and tannin/crop, and (2) less shrivelling in the berries. An adverse effect is a lesser volume of juice/crop. There is also, as demonstrated in Table 1-7, a lower overall crop. However, if quality is the main objective and it is economically viable to have the lower yields, then the best wines will come from these selections.

Table 1-9. Pinot Noir Clonal Selections, Wine Analyses, Six-Year Averages (1980-1985).

Selection	Total Acidity H_2Ta g/L	pH	Ethanol % v/v	Potassium mg/L	Total Phenol[1] mg/L	Color O.D. 420/520 nm
A	7.8	3.31	13.4	853	1203	1.69/2.96
B	8.4	3.15	13.0	758	1417	1.73/3.40
C	8.1	3.19	13.0	779	1358	1.73/3.20
D	8.1	3.26	13.0	800	1213	1.58/2.96
E	7.7	3.36	14.2	808	1513	2.48/4.58
F	8.5	3.05	12.5	674	1114	1.51/2.95
G	7.9	3.27	13.7	759	1415	2.29/4.30
H	8.0	3.23	12.3	785	1257	1.66/3.07
J	7.8	3.26	13.4	840	1576	1.76/2.96
K	7.9	3.21	13.6	782	1425	1.72/3.29
L	7.8	3.32	13.4	844	1397	1.72/3.05
M	7.8	3.32	13.9	891	1295	1.90/3.37
N	8.0	3.22	12.8	778	1376	1.69/3.06
O	7.9	3.37	13.7	915	1392	1.78/3.04
P	8.0	3.23	13.7	800	1478	2.30/4.32
R	7.8	3.27	13.6	815	1360	1.86/3.31
S	8.1	3.22	13.0	843	1282	1.67/2.96
T	8.5	3.10	13.2	765	1414	2.10/4.20
V	8.1	3.20	13.4	834	1408	1.84/3.39
Z	7.8	3.26	13.4	817	1436	1.80/3.19

[1]As gallic acid equivalents.

Efforts of excellent winemakers to make consistently superior wine from Pinot noir have met with moderate but ever-increasing success. This variety is less rewarding for the amateur than for the professional.

A companion variety, Meunier, has similar characteristics but

Table 1-10. Berry Size Relationships.

	Mean berry wt. (g)		
Variables	1.59	1.34	0.96
Mean pulp wt/berry (g)	1.24	0.99	0.64
Mean skin wt/berry (g)	0.28	0.29	0.27
Pulp/skin ratio	4.43	3.41	2.37
Relative transpiration rate %	100	66	33

even less flavor and color. It is suited for champagne if harvested early enough.

Chardonnay, the premier white variety of Burgundy, has been greatly improved by recent clonal selection work in California. Before this work, the variety had low crop yields due to poor fruit set. This distinctive variety, when grown in a cool area, makes a wine of excellent quality. Chardonnay aroma is not defined by any one or any family of compounds. Simpson and Miller (1984) identified many of the volatiles. The only ones recognized as important and associated with Chardonnay aroma were the fermentation alcohols and esters, acetic acid and damascenone. They noted these were common to many varieties.

When Chardonnay is fermented in wooden barrels or aged in wood, that contributes to the aroma. Also, it is common practice to subject at least part of the wine to malolactic fermentation. This also adds complexity. Skin contact and its effects are discussed in Chapter 2.

Petite Syrah, another French variety, has a slight varietal aroma sometimes described as smokey. It is a good standard variety and yields wines of good body from either moderately warm to cool areas. Its intense color makes it a good blending variety. The fruit has a tendency to sunburn and raisin under maturing conditions that are too warm. Be sure to obtain the true French strain (now called Syrah or Shiraz) and not the more commonly used one whose true name is Durif. There are significant yield differences between the higher yielding Durif and the Syrah.

Zinfandel apparently comes from Europe. This variety, a good producer, has a distinctive raspberry-like aroma when grown in the cool, but not cold, regions. The tight clusters make it more susceptible to mold and rot. It has a unique varietal aroma but, when aged for an extended period, resembles a Cabernet of similar age. One common use is to make a "blush" wine as this type has good body and makes a very rich, lightly colored wine, easy to drink.

Barbera, a variety from northern Italy, makes a fruity tart wine that can either be used by itself or as blending material for less acid wines. It produces its best wine when grown in a moderately warm area. In areas that are too cool, astringency and acid become too high. It has not tested that well in California.

Muscat-type wines consist of Gewürztraminer, White Riesling (Johannesburger Riesling), Muscat blanc (Muscat Frontignan), Orange Muscat and a host of muscat-flavored table grape varieties. The latter should not be planted for winemaking because of their low acidity. The muscat aroma comes from the presence of compounds called terpenes and related compounds.

A great deal of work in the last few years has delineated those terpenes. The concentrations of the various terpenes vary from one muscat variety to another, giving the subtle difference that can be found in the aromas. A review by Williams et al. (1987) notes that 35 different monoterpenes have been identified in grape juice and wines. Some of the odorless glycosides of terpenols and polyhydroxylated alcohols and polyols can be converted into aroma compounds.

The search for enzymes to hydrolyse these glycosides is underway by several laboratories. Most of the glycosidases are inhibited by glucose. There are more of the glycosides in the juice and wine than there are free terpenes. The glycosides are odorless and essentially tasteless. A list of some of the major terpenes found and their formulas are given in Table 1-11. There are some natural glycosidases in yeast which will cause some increases in the terpene aromas during fermentation and aging. Some skin contact or heating will also cause small increases in the terpenes and terpene glycosides.

Table 1-11. Some Common Terpenes and Their Formulas Found in Muscat Type Grapes.

Linalool

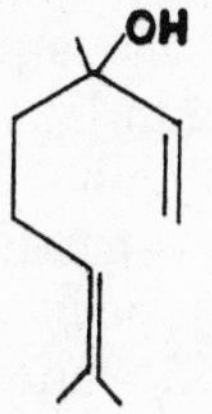

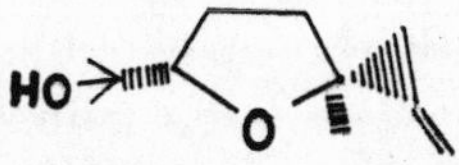

cis-Furan linalool oxide

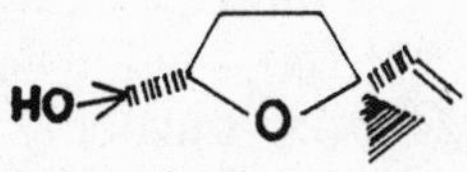

trans-pyran linalool oxide

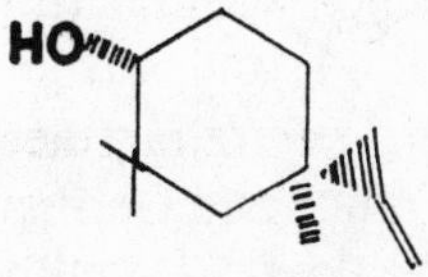

Diendiol

cis-pyran linalool oxide

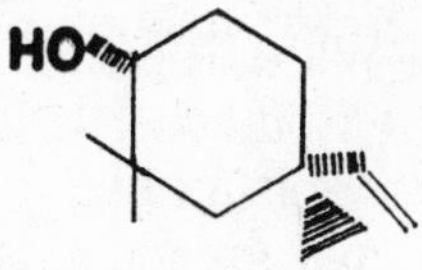

Geraniol

α-Terpeniol

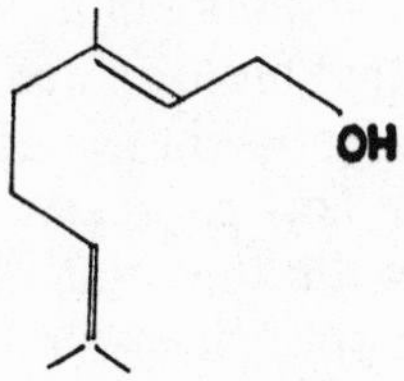

Nerol

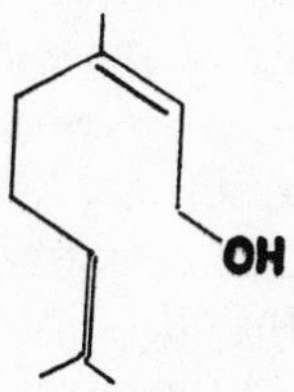

trans-Furan linalool oxide

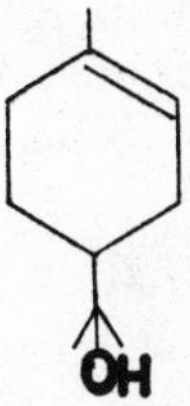

Cordonnier et al. (1986) reviewed past work on terpenes and compared non-muscat to muscat varieties. They speculated on methods to hydrolyse the precursor glycosides to the aroma-producing terpenes. Marais (1987) looked at the terpenes in the Gewürztraminer cultivar and found trans-furan linalool, alpha-terpeneol, citronellol and diendiol-1 increased about 10% from 22° to 24°Brix. Linalool, nerol, geraniol and trans-geranic acid decreased by 10 to 50%. Nerol, geraniol and trans-geranic acid were there in larger relative amounts of 29, 123 and 52 while others were less than 10 at 23°Brix. The actual value of total potential volatile terpenes was determined by the method of Dimitriadis and Williams (1984) and was about 1 mg/L.

Gewürztraminer comes from the cool Alsace region between France and Germany. It makes a distinctive wine with a fruity muscat flavor. It yields wines of a flat and coarse nature when grown in warmer areas. It ripens very early.

White Riesling is one of the quality wine varieties of Germany. There it consistently yields wines of sound and distinctive character. It is a suitable variety for cold winters but it does poorly when grown in moderately warm areas that are irrigated or receive summer rains. Because of its relatively soft skin it is easily attacked by molds, including *Botrytis cinerea*. This particular mold causes the grapes to be dehydrated and gives the flavorful "Auslase" of the Rhine. This infection depends on specific natural climatic conditions that do not occur every year. It has become less popular as a table wine variety with planting constantly decreasing in California.

Orange Muscat and Muscat blanc are poor producers but have some favorable attributes for making sweet table wine. They have little or none of the harshness and bitterness associated with wines made from their table grape cousin, Muscat of Alexandria. These two varieties ripen sufficiently in the cooler areas and both will raisin if not picked at the proper time. Do not plan to make an acceptable dry wine from these two varieties. Symphony is a newer *vinifera* hybrid that makes an excellent muscat dry wine or slightly sweet wine when grown in the cooler areas. It lacks much of the bitterness associated with one of its muscat parents, Muscat of Alexandria, and has good acidity and color.

Other *vinifera* varieties which occasionally are of interest to the amateur winemaker are listed in Table 1-12 with their major weakness noted. Some of these varieties can be used with success but usually they should be avoided.

HARVESTING

When to Pick

Grapes should only be harvested when the winemaker can handle them immediately. Pick only on the day they are to be crushed. By early sampling, date of picking can be at least partially anticipated

Table 1-12. A Limited Number of Varieties Not Recommended for High Quality Table Wine.

Variety	Some Difficiencies
Alicante Bouschet	Poor color stability
Aligote	Ordinary quality wine
Burger	Poor quality wine, low acid
Carignane	Ordinary quality wine
Clairette blanche	Ordinary quality wine, low acid
Folle blanche	Fails to ripen properly
Furmint	Late ripener, ordinary quality
Grey Riesling	Difficult to harvest at optimum maturity
Green Hungarian	Poor quality wines--no character
Grignolino	High tannin, low color, different
Limberger	Ordinary wine
Palomino	Subject to rot, low acid, poor table wine, good for sherry
Pinot St. George	High pH wines often of poor quality
St. Emilion	Course wines, ordinary quality
Sauvignon vert	Thin skinned, subject to rot, very low acid
Sylvaner	Subject to rot and mildew, low juice yield
Tinta cao	Port variety, very high pH, poor quality wine
Tinta Madiera	Port variety (excellent for port) high pH and odd flavor for dry red wine

(Figure 1-8). If the grapes are in excellent condition with no broken skins and are kept cool, a one-day delay is tolerable. Also, be sure to pick only ripe clusters. This is especially true of Zinfandel which has a tendency to set a second crop that will only be partially mature. If they are mixed with the normally ripe fruit, the average sugar content is lowered and the acidity raised. This gives a wine that can be thin (low in alcohol) and too tart for a red wine. The exception for this is when the grapes are to be made as a "blush" wine, then the higher acid second crop improves the balance of the wine.

Several tools are used as picking aids. Shears are the easiest and safest to use. Picking knives are apt to cause damage to inexperienced hands. Large plastic dishpans or aluminum picking pans are easy to pick into and are less of a chore to pack out of the vineyard than larger containers. Handle the grapes with care.

If white grapes are picked during a hot day, either chill the fruit before crushing or chill the juice as soon as it is extracted. Cooling the grapes reduces the enzymatic browning, allows better control of the fermentation and produces a better wine.

Picking at night is also a good answer to having cool fruit. If no SO_2 is desired in the juice, then being cool slows any bacterial or wild yeast growth. Normally it is desirable to have some skin contact for certain varieties; this is best done while the must is cool. Picking at night results in fruit at 55-65°F (12.8-18.3°C) and the cooler condition improves the pickers' production. There are more injuries reported in night picking.

Mechanical Harvesting

Grape mechanical harvesters (Figure 1-9) have been consistently improved. Over the row machines can pick most varieties. Some varieties mechanically pick better than others. A list of varieties and some of their mechanical harvester picking characteristics are given in Table 1-4. White varieties should be picked either at night or on cool days. The mechanically harvested grapes are much more subject to damage than hand-harvested grapes. They must be handled as rapidly as possible. Some operations add sulfur dioxide during the mechanical harvesting. Others have the grapes crushed in the

FIGURE 1-9. A two row over-the-row harvester that can be modified for other vineyard chores.

field immediately after mechanical harvesting, and rapidly transport the crushed fruit to the winery for pressing to minimize changes. Cabernet Sauvignon grapes are the easiest to harvest mechanically. French Colombard grapes are one of the most difficult as they leave much of the juice on the vine. Expensive grapes, except for Cabernet Sauvignon, probably should be hand harvested.

Fruit Quality

Some wineries sort their varietal grapes to get rid of the rotten fruit or second crop. While this is very costly, it does allow for the ultimate quality to be achieved. Most wineries of any size would have to be very selective of how much effort was spent in this manner.

Dealing with the quality of the fruit is always very subjective. The usual way in most areas is to pay basically so much a ton and specify a °Brix range which is acceptable. Others have specified, in addition, a maximum allowable MOG (material other than grape), usually in the range of 1-2% wt/wt. This is a rather difficult measurement. It is usually done by sampling a load, then taking subsamples and finally separating and measuring the leaves, twigs, etc., in the subsample. Another criteria is mold and rot. This is very difficult to estimate properly. Again the subsamples can be visually searched for rotten and moldy fruit. However, with mechanically harvested fruit this becomes extremely difficult. Recent efforts have been made to develop monoclonal antibodies so the subsamples can essentially be typed for the kind. These antibodies react specifically with certain components unique to the mold. The reactants can be measured by various techniques. Looking at such chemical components as acetic acid, alcohol or glycerol is very impractical. Any yeast present will make far more than could be attributed to any mold or rot activity in cases of delayed delivery. While alcohol, ethyl acetate and glycerol relate to the fruit quality, they cannot accurately be used to predict mold and rot.

Other ways to be assured the fruit is received that makes the desired kind of wine is to keep lots separate from the various sources and evaluate the wine quality with the vineyard (or, in addition, various treatments) in mind. If those grapes make the desired kind of wine, then continue to buy them and pay accordingly. If the wines from a specific vineyard aren't acceptable, simply refuse to buy the grapes anymore. This is the only way, if grapes are purchased, that wine quality will be optimized. It is highly desirable to grow a portion of your own grapes. Buying a percentage of grapes does allow some economic flexibility that is important and necessary. Figure 1-10 shows a winery tank system for doing lots of 50-100 gal for test purposes. In most wineries, more than one vineyard picking usually is combined into one tank.

Botrytis cinerea on the fruit is common in areas where summer rain is a problem. One of the major chemical changes that occurs from a *Botrytis* infection of the grape is oxidation of the sugars. Table 1-13 shows some changes occurring with gluconic acid within the berry (Pineau, 1987).

Other acids formed from oxidation of sugars are glucuronic, galacturonic (from pectins) and mucic. The sugars are slightly depleted and nitrogen can decrease by 80% (making the use of diammonium phosphate for yeast nutrition recommended). Alditols, glycerol and acetic acid increase. Ravji et al. (1988) verified that glycerol can be formed even under sterile conditions. The amounts vary from grape to grape and mold to mold. The dehydration of the fruit causes the concentration of all the components to levels well beyond normal.

The *Botrytis* infection of berries causes the diminution of positive aroma factors (Bayonove, 1989). The terpenols are reduced to low levels. The favorable esters, after and during fermentation, are hydrolysed rapidly by the esterases from the *Botrytis*.

BUYING GRAPES (HOME WINEMAKER)

If there are wine grape vineyards reasonably close, a negotiated purchase from the owner is a very satisfactory way to obtain your fruit. Sometimes, for home winemakers he may agree to let you pick your own fruit. This will allow you the feeling of being a complete vintner. However, this feeling is seldom worth the aches and pains. Before contacting the farmer, it would be wise to inquire about and determine the going commercial price of the variety you wish to buy. This will give you a negotiating point.

Grapes can be obtained from brokers who purchase grapes from a

FIGURE 1-10. A commercial winery pilot plant showing small fermenters to test individual plots or vineyards (Nederburg Winery, Paarl, South Africa).

Table 1-13. **The Amounts of Gluconic Acid Formed by the Various Degrees of *Botrytis cinerea* Infections of the Berry Surface.**

Surface infected, %	Gluconic acid in the berry, mg/L
0	29
20	121
40	408
60	420
100	1240

local vineyard. They package them and ship them to various areas within the United States. The quality of grapes obtained from this source can vary greatly. Some brokers ship fruit of a questionable variety, do not take proper precautions in shipping and obtain the fruit from less desirable growing areas. Before buying grapes in this manner, you should obtain the answers to the following questions:

1. What variety of grapes are being sold?
2. In what area (county) of what state were they grown?
3. Have the grapes been gassed with sulfur dioxide, held and shipped in cold storage?
4. What will the sugar content and acidity be?
5. Can I obtain a refund if the grapes arrive in a deteriorated condition?

There are few wine grapes than can be successfully shipped. For this purpose, the best variety that makes quality wine is Cabernet Sauvignon. Some others are Sauvignon blanc and Alicante Bouschet, with the latter making only mediocre wine. The main criteria are loose clusters and tough-skinned berries that are held tightly to the stem. Grapes grown in the cooler regions on non-irrigated soils ship better than those from warmer irrigated vineyards. Grapes will mold easily. Moldy grapes should not be used for wine. Shipped grapes should be closely inspected and one should be fairly critical of their condition on arrival.

RAISINS (HOME WINEMAKER)

The use of raisins for winemaking is a practice that should be avoided. Besides the expense of using raisins, the quality of wine resulting from their fermentation is very poor. The odor and flavor of the raisins remain in the wine. The color of the wines is dark and murky. Raisins are made primarily from table grapes (Thompson Seedless and occasionally Muscat of Alexandria) which, at best, make less than ordinary table wine.

CONCENTRATE (HOME WINEMAKER AND COMMERCIAL WINEMAKER)

Concentrate is probably the most popular source of fermenting material for home winemakers living in a non-grape-growing area. Grape juice concentrate is usually made from juice obtained from table grapes or the less expensive wine grapes grown in the warm interior valleys of California. Most concentrate is produced to be used to sweeten canned fruit or for other sweetening purposes. Unless a wine variety of grape was designated, the chances that the concentrate was made to be reconstituted for fermentation are small. How the concentrate is treated after it is produced is sometimes as important as the production techniques. Some useful conditions follow. Store with sufficient sulfur dioxide to prevent browning, oxidation and yeast growth. Keep at moderate storage temperatures. Tartrate stabilize (remove excess potassium bitartrate salts) before concentration, or the concentration will be very cloudy.

Time is the enemy of the concentrate, especially when stored in metal or plastic containers. Metal containers are easily attacked by the grape acids. When this happens, the concentrate cannot be used for wine. Polyethylene containers allow transmission of oxygen into the concentrate. Red concentrate is made from grapes that have been heated. It is difficult to obtain a stable juice even if the good wine grape varieties are used. It is common practice to use hybrid grapes that have colored juice for red concentrate. These do not make quality table wine. The most common hybrid color variety is Rubired or Salvador. If good wine varieties of grapes are used and they are treated properly before, during and after concentration, reasonably good wine can result. It is difficult to maintain the delicate varietal characteristics, but clean fruity wines of an acceptable quality can be made. Some questions you would ask are:

1. Is the white grape concentrate light in color?
2. Is the red grape concentrate red in color rather than brown?
3. Is the shipping container glass?
4. Does the label give the variety of grape used?
5. Does the label give the area in which the grapes were grown?

If the answers to the above questions are yes, then the concentrate is probably acceptable. It can be reconstituted and fermented with moderate success. Don't expect high quality wines.

Commercial wineries seldom use concentrate for refermentation but do use it to sweeten table wines. The above advice, in general, holds for both commercial and home winemakers.

RECTIFIED GRAPE JUICE CONCENTRATE

The making and use of a rectified grape juice concentrate is a question that has been discussed in some technical detail (Dupuy, 1985). The question that surfaces with governmental regulation agencies is "When is a fruit juice concentrate no longer representative of the fruit it was made from?" There are several processes applied to grape juices (and other juices) before concentration. These will stabilize the concentrated material to any further changes when it is stored and used later. These processes are:

1. Cold stabilization to remove excess potassium bitartrate.
2. Pectic enzyme treatment to allow easier filtration and to clarify the juice.
3. Flash pasteurization or bentonite treatment to stabilize against protein precipitation.
4. H^+ cation exchange to remove amino acids and minerals and lower pH to prevent microbial activity.
5. HCO_3^- anion exchange to remove acids and phenols to prevent browning during concentration, storage and later use.

After these treatments, the juice is essentially sugar and water. Any volatiles left are removed during concentration. The first three steps are those taken to make normal grape juice concentrate and are essential to make a reasonable product. With the ion-exchange treatments, the changes are much more dramatic. The concentrate is much easier to handle and use but a poorer representation of a grape product.

JUICE STORAGE

Grape juice of wine grapes, if from sound fruit, can be kept under cold conditions with minimum damage to the juice. The juice should be kept at a temperature that is close to its freezing point (−3°C). The juice, if air is excluded, will survive well. It can be warmed up, yeasted and fermented to give fruity non-varietal or even varietal wines of good quality. Sufficient cooling must be available. Any laxity with the cooling and a yeast infection starts. It can seldom be stopped without a maximum effort. Add 50-100 mg/L sulfur dioxide (SO_2), treat with pectic enzyme and hold for 4-6 hours to allow close filtration, then rapidly chill to the holding temperature. This juice can then be held satisfactorily for many months. Air must be excluded or molds will grown on the surface of the juice. While mold will not grow on a wine surface, it grows very well on juice, even at −3°C, if air is present. Juice held can be fermented later in this manner. This results in fruity wine to blend with earlier fermented wines which have lost their original fermentation bouquet.

The use of the so-called "Brimstone" process of grape storage is falling into disfavor. In this process, the juice is kept with the addition of 1000-2000 mg/L of SO_2. Then the juice is recovered by running it through a machine to pull off the SO_2 with heat and under vacuum. The SO_2 is trapped as the calcium salt. The juice oxidizes readily and ferments poorly. The vitamin thiamine, essential for good fermentation, is destroyed by the treatment. In addition, the high level of SO_2 is very damaging to the stainless steel tanks and equipment. The idea was to have a supply of juice to ferment into fresh fruity wines year around. Because of the above factors and the rather poor wine quality that results, it is not recommended.

Chapter 2

Grape Processing

After the grapes are picked and transported to the winery, prompt and proper handling is important. If the grapes were picked at a cool temperature, they can be processed immediately. On the other hand, if the white grapes are warm-hot (over 25°C), they should be cooled to at least 20°C before crushing, if practical. Red grapes can be crushed while still warm as they will ferment at a higher temperature. The beginning temperature of the grapes is less important to the final product. The fermentation temperature must be controlled. Be aware of the need for refrigeration.

EQUIPMENT

Stainless Steel

All modern table wine wineries now use stainless steel tanks and most are jacketed to allow temperature control (see Figures 2-1, 2-2). The stainless steel used for the body of the tanks is 304. It consists of Fe, 73.44%; Cr, 18%; Ni, 8.5%; and C, 0.06%. When the joints are welded, 316 stainless steel is used (Fe, 68.24%; Cr, 17%; Ni, 12.5%; Mo, 2.2%; and C, 0.06%). Three hundred sixteen is also used for the roof of the tanks that may be exposed to SO_2. The stainless steel is pacified by pretreatment with nitric acid. The chromium oxide that forms is the actual protecting coat. If it is reduced by SO_2 or some other reducing agent, the molybdenum in the 316 stainless steel helps reform the chromium oxide. Chlorine can cause pitting. The size of tanks can vary from 500 gal to 150,000 gal or more. Previous to the use of stainless steel tanks, soft iron tanks coated with epoxy resin were used. These are inert

FIGURE 2-1. Large unjacketed wine storage tanks with awning to protect from the heat of the sun.

like stainless steel, but the coating can be damaged and is costly to repair.

While all equipment does not need to be stainless steel, the more that is stainless steel, the less the chance for metal pickup. Further consideration of this will be covered under wine stability.

FIGURE 2-2. Smaller jacketed fermentation and storage tank using direct expansion coolant for refrigeration.

Cooling

The main system of cooling is to use glycol to circulate through the jackets. Some use ammonia direct expansion systems. Some fermentors are not jacketed. In these tanks, the wine is pumped through a heat exchanger and back into the tank. Many wineries of medium to large size have computer programs to control temperature. These monitor the tank temperatures and regulate the refrigeration to maintain desired conditions.

Movement of Grapes

The movement of grapes before crushing is usually done by screw conveyors which move the fruit dumped from the gondola into a slanted receiver (Figure 2-3). The screw at the bottom moves the fruit with minimum harm to the stemmer-crusher. After crushing and stemming (Figure 2-4), the fruit goes into either a drainer, large holding tank or with red grapes, into the fermentor. The crushed grapes are usually moved by a Mono® pump (Figure 2-5) or a piston pump. These pumps can handle material with very low liquid content and not become clogged.

Maceration carbonique requires that the bunches be intact and relatively undamaged when they are put into the closed tank for fermentation. To cover the grapes with more juice or partially crush the grape is not doing the classic "maceration carbonique." See Chapter 3 for details.

Cleaning

All equipment should always be clean and suitable for fermentation. If it is not, then problems such as off-odor and -flavors result. Water is the best solvent for this purpose. The use of high-pressure jets of hot water will remove most contaminants. Beyond that, the use of chelating agents such as ethylenediamine tetraacetate (EDTA), nitrioltriacetate, sodium tripolyphosphate and others will remove metals by complexing with them. They all are more effective in an alkaline medium. After washing the equipment, a tank should be well rinsed with a citric acid solution to neutralize the

FIGURE 2-3. Dumping bin for grape transport trucks or gondolas.

FIGURE 2-4. Dumping bin feed screw moving grapes into the crusher-stemmer.

caustic. Depending on the conditions of the water used, the strength of caustic solutions used to remove tartrates from tanks should be between 1-2%. The harder or colder the water, the more caustic soda is needed. Surfactants can be either the ionic or non-ionic type. They emulsify and wet the material, allowing the material to float off and be removed. Most sanitizing agents used furnish a source of chlorine. Contact time of 3 min. at 200 mg/L of chlorine is sufficient to sanitize tanks or lines. Stainless steel can be attacked by chlorine. More damage can result if water containing high amounts of salt is

FIGURE 2-5. Mono® pump used to move crushed grapes.

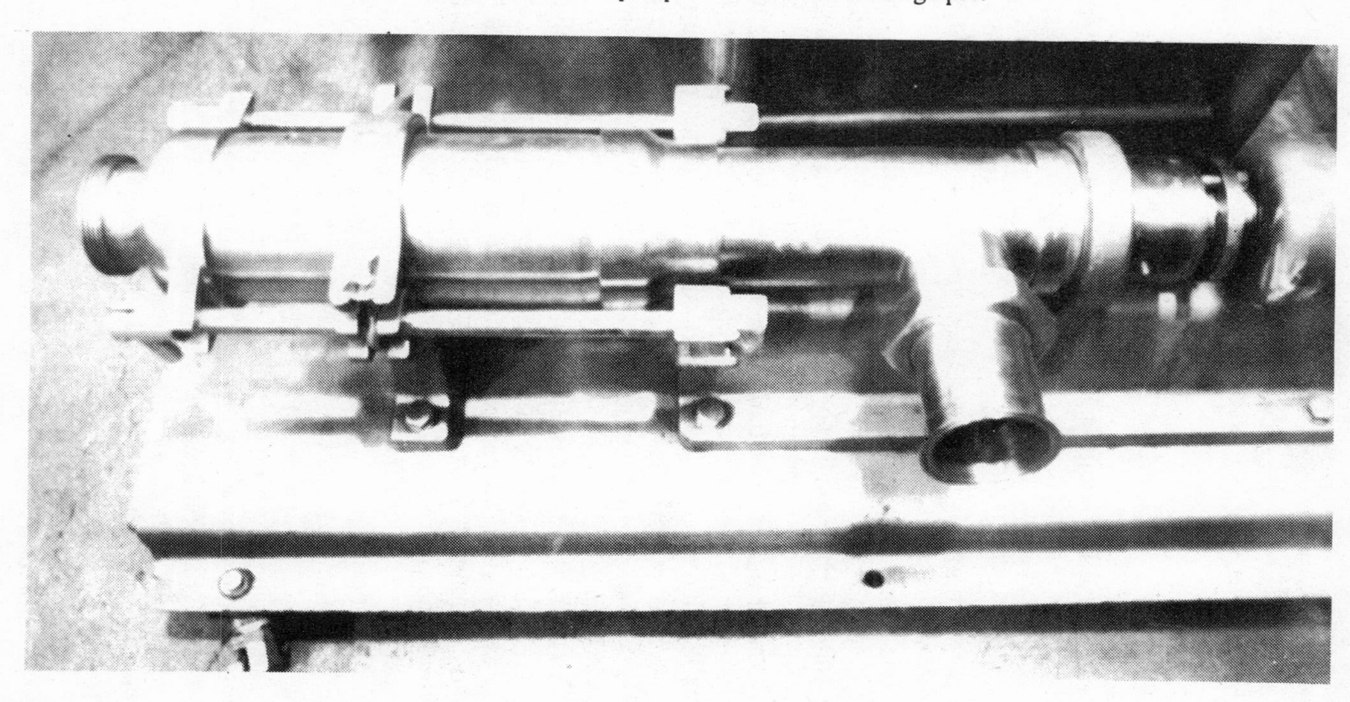

used. The contact time limitation should be observed. See Chapter 9 for some pertinent analyses.

STEMMING AND CRUSHING

Normally, both white and red grapes are crushed, stemmed and the stems removed. Occasionally, to facilitate drainage during pressing, the stems are left in contact with the crushed white grapes. However, if the stems, from most varieties, are left in contact with the crushed grapes for an extended time, a "stemmy" off-character will result in the wine.

The stemming can be done by several methods. The older type stemmer-crusher has a rapidly spinning blade at the mouth of the crusher which immediately impacts the cluster and destems it. The stems and berries are then moved along an inner perforated drum by a series of blades. The berries are thrown out through the perforations into a slower moving outside drum with perforations of smaller diameter. A large sweeping blade pushes the berries out through the perforations, crushing the remaining whole berries. The stems remaining in the inner drum are moved to the end of the drum and discharged.

Another type of crusher-stemmer (Figure 2-6) is a roller type. The grape clusters drop onto the adjustable rollers and are crushed. The stems and berries drop into a perforated drum with several round spinning blades. These are arranged to move the berries out through the perforations and the stems out the end.

The object of crushing is to open up every grape so the juice can be easily drained with a minimum of damage to the grape skin. If the berries are left intact, they may go through pressing unchanged and the juice will be lost. The chemical composition of the juice can also be changed during crushing. Table 2-1 demonstrates the differences in the content of the juice produced from the same grapes handled by different type of crushings.

These differences in the content of the juice result from the amount of shearing or tearing force applied to the berry. Berries which are less turgid (not as firm) or more shrivelled (raisined) are more difficult to break open. Applying too much force causes the skins of the berries to break into many fragments. This action (mac-

FIGURE 2-6. A standard roller crusher with stem separation cylinder.

Table 2-1. Effect of Crushing Type on the Juice Composition. (Average of nine different white grape samples).

Crushing juice	°Brix	Total acidity g H_2Ta/L	pH	Total nitrogen mg/L
A. Free run juice[1]	22.3	7.30	3.52	664.0
B. Fruit slurried in a Waring blender (without stems)[1]	22.6	7.32	3.79	761.0
C. Whole berry slurried and fermented[2]	-	-	-	1859.00

From Ough (1969).

[1]Centrifuged juices prior to analysis, seeds not crushed.
[2]Whole grape (no stems), from same number of berries to give B.

eration of the skin tissues) stimulates enzyme activity and may cause undesirable reactions. In addition, many of the grape cells are ruptured, releasing their contents into the juice. These strongly buffered cellular juices elevate the pH of the juice and resulting wine. If the grape skins are highly macerated, they become a problem in the clarification of the juice. Care must also be taken to keep the seeds intact. When the protective outer shell of the seed is broken, the high levels of phenolic material that the seed contains will impart a bitter taste to the wine.

JUICE SEPARATION

Red Varieties

Immediate separation of the juice of red wine from the skins is seldom necessary. Only if one wishes to make a white wine or a "blush" type from red grapes is this done. For white champagne from Pinot noir, it is necessary to make this separation rapidly. A minimum of damage to the skins results. In *vinifera* grapes, the color is in the skin. Early picking of the Pinot noir grapes when the °Brix is no more than 18°B to 20°B is important. The champagne stock should have no more than 11% alcohol. Early picking will also allow less color to be extracted during pressing. The most common, but inefficient, practice for obtaining white juice from red grapes is to press the whole cluster gently without prior crushing.

Any type of press is satisfactory, including a simple hydraulic or hand-operated basket press. The juice coming from the press should be carefully monitored. When the juice becomes a definite pink, this juice should be kept separate from the previous juice. The remainder of the juice extracted can be fermented into pink champagne. Considerable juice will remain in the grapes when no more juice can be extracted by gentle pressing. For Pinot noir, this material can be fermented even though the stems will remain with the grapes. With Pinot noir, the stems impart a pleasant peppery flavor. If desired, the grapes can be destemmed before fermentation.

Rosé wines are usually made from varieties having stable color. Cabernet Sauvignon, Zinfandel and a French variety called Napa Gamay are some of the most commonly used. In the past, Grenache was used but its tendency to turn brown rather easily caused it to fall into disfavor. Now with the popularity of blush wines, Grenache is again being used for that purpose. There is no firm rule for the separation of juice from the skins in rosé winemaking. Overnight skin and juice contact, with a moderate amount of fermentation (a drop of 1-5°Brix), is sufficient to allow the necessary color pigments (anthocyanins) to be extracted. This is a judgement call and will vary from lot to lot. The blush wines are lighter than rosés.

After the blush or rosé stock has been separated, it should be treated as a white juice. It should be protected from oxidation. The pigments are partially stabilized by the addition of SO_2. While addition of SO_2 will initially bleach the anthocyanin pigments, it also stabilizes them from oxidation to brown polyphenols. After fermentation there is a slow release of these pigments. The SO_2 equilibriums proceed (Figure 2-7) as the available free SO_2 is oxidized.

With white champagne stock from red grapes, it is advantageous not to add SO_2. Let the oxidation of the pigments and phenols proceed by allowing some air contact. The monomers will condense.

FIGURE 2-7. Effect of pH and HSO_3^- on the various anthocyanin equilibriums.

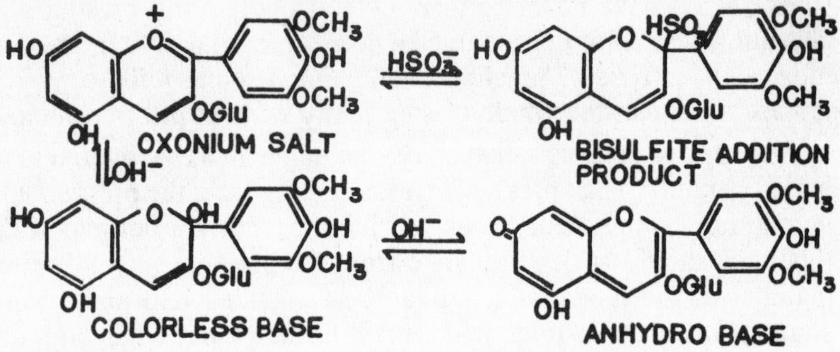

After fermentation, initial fining and filtration, the wine will be as light in color as by any other treatment.

The first SO_2 treatment should be when the wine is racked off the lees. After this point, non-enzymatic oxidation can proceed. The SO_2 acts as the reducing agent to scavenge the oxidizing components formed. For standard red wines, crushing and stemming is the only physical pretreatment before fermentation.

When pressing red grapes after or during fermentation, the degree of press should be monitored. The hard press juice is kept separate for blending back later to achieve the desired flavor and color. Hard pressing of the skins causes disruption of the cells and more phenols and other components to be released. It may be desirous to use this material later, and should be handled as carefully as the low- and medium-pressed material. Special fining of the wines made from the heavy pressings will reduce the phenolic content.

Thermovinification

Thermovinification is seldom used for color extraction in most quality table wine operations. This method involves the rapid heating and cooling of crushed grapes. The heat kills the grape cells and allows the color pigments to be released into the juice. It also allows for rapid juice separation. While the wine is not harmed appreciably by this treatment, it does impart different flavors and aromas into the juice. The resultant wine is of a different character and can be easily differentiated from wine made in the conventional manner. The consensus of most winemakers is they prefer the conventional methods. In certain areas (France and other areas with summer rain), *Botrytis cinerea* mold (grey mold) is a very serious problem. The laccase enzyme from this mold will oxidize the red pigments in the wine to brown polymers. To prevent this, grapes are heated before vinification to inactivate this enzyme. There is also a polyphenol oxidase enzyme present normally in grapes. This enzyme is a tyrosinase and is inhibited by minor amounts of sulfur dioxide, if that is desirable in the juice. It is not as destructive as the laccase from the *Botrytis*. The laccase is not inhibited effectively by SO_2.

Polyphenol Oxidases

The variations in the relative oxidative properties of tyrosinase (grape enzyme) and laccase (*Botrytis* enzyme) were compared by Ribéreau-Gayon (1977). He showed that the laccase effectively oxidized most of the phenols present in grape juice. Tyrosinase was effective only against several of the phenols (catechins, pyrocatechol, caffeic acid, chlorogenic acid and protcatechic acid). The rate of uptake of O_2 did not vary significantly between normal Sémillon juice and that from *Botrytis*-infected fruit. White and Ough (1973) found that 25 to 35 mg/L of SO_2 were sufficient to inhibit the tyrosinase in most juices. The exception was a Pinot blanc juice that took more than 100 mg/L to inhibit the oxygen uptake completely. As expected, temperatures slow the enzymatic activity but it still proceeds at 0°C. The rate of oxygen uptake is a rough measure of the polyphenol oxidase enzyme activity (tyrosinase) and the amounts of oxidizable substrate present. The uptake rate in some California grapes is shown in Table 2-2. Both factors are contributing to the rates. The oxidizability of the individual lots is an index that will vary from variety to variety and vineyard to vineyard.

Colloid formation in wines from *Botrytis*-infected grapes is common. Pinot noir grapes treated with thermovinification methods to deactivate the laccase polyphenolase were investigated by Villettaz (1988). He found little differences between the colloid content of these wines made by thermovinification or normal fermentation. The total colloid amount was 324 mg/L and 914 mg/L for the juice

Table 2-2. Relative Oxygen Utilization Rate Constant for Various Grape Varieties.

Variety	K	Variety	K
Pinot blanc	0.948	Chenin blanc	0.036
Grenache	0.358	Flora	0.015
White Riesling	0.336	Semillon	0.013
Pinot blanc	0.326	La Rehiena	0.011
Emerald Riesling	0.326	French Colombard	0.006
White Riesling	0.271	Grey Riesling	0.006
Helena	0.236	Muscat of Alexandria	0.004
White Riesling	0.182	Semillon	0.013
White Riesling	0.092	Sauvignon blanc	0.002
Ruby Cabernet	0.059	Flora	0.000
Clairette blanche	0.045	Merlot	0.000

From White and Ough (1973)

and wine, respectively. The alcohol-insoluble neutral fraction of polysaccharides in the juice and wines were 77% and 87% of 252 and 722 mg/L colloids, respectively. The pectins were 20% and 8%, and the proteins 3% and 5%. In the alcohol-soluble portion, 72 mg/L for the juice and 192 mg/L for the wine, the juice contained 100% neutral polysaccharides and the wine 94%. The residual 6% in the wine fraction was protein. These colloids were degradable by β-gluconase enzymes. The sugar content in percent is given in Table 2-3.

White Varieties

Traditionally, white varieties for table wines were separated from the skins and seeds as rapidly as was feasible. They were fermented at a temperature that was convenient to keep the vintage moving along. Very little care was exercised about the fruit temperature,

Table 2-3. Percent Sugar Composition of Neutral Polysaccharide Colloids from Pinot noir Thermovinified Juice and Wine.

	Insoluble fraction in 83% ethanol		Soluble fraction in 83% ethanol	
Sugar	Juice	Wine	Juice	Wine
Rhamnose	4.1	6.1	1.7	7.7
Arabanose	12.9	20.8	65.5	62.5
Mannose	19.9	34.3	14.7	10.4
Glucose	12.6	7.3	10.6	9.8
Galactose	48.5	30.3	7.4	10.1
Fucose	2.9	2.2		

From Villettaz (1988)

skin contact, use of SO_2 or severity of press. All of these have become of intense interest to the modern winemaker who wishes to make the best quality product possible.

Temperature

Grapes are close to the ambient air temperature. Cooling the grapes directly by forced air chilling systems is difficult, as most grapes are now harvested into trucks or gondolas. There is no way to circulate the cold air effectively. Cooling the juice can be achieved during settling and is an option for those not having extended skin contact. Wineries can run the white crushed grapes through heat exchanges before pressing and lower everything to the desired temperature. This is especially useful for white grapes that will have a long skin-juice contact time. It is also useful for red grapes being used for white or blush wines as it decreases color extraction. Some wineries also use cooling for red grapes if they come in at excessively high temperatures so the fermentation can be more easily controlled.

Thermal Processing of White Juice

According to Heimann (1977), all clarified white juices in the largest German winery are heated to 87°C for 2 min then cooled to 15°C before yeast inoculation. This stabilizes the future wine to protein cloud. He also found experimental treatments of heating musts, while treated with pectic enzymes, aids in increasing pressing yields. The optimum pressing time with heating at 45°C and enzyme for 2 hrs was 34 min compared to 85 min for standard pressing. Effective juice yield improvement can be achieved at 20°C for four-hour treatment (Ough and Crowell, 1979) (see pectic enzyme section).

SKIN CONTACT

In white varieties, the rapid separation of the juice from the skins and seeds is important as there are significant amounts of tannin-like material in the skins. Extended contact with the skins allows these materials to be extracted and remain in the finished wine. These compounds are kept at a minimum by rapid separation of the juice from the skin. Any skin-juice contact (after crushing) exceeding 12 hours at normal cellar temperature may be harmful to the resulting wine. Any time over four hours causes noticeable wine quality changes with most varieties. These changes are in color, taste and aroma. At elevated temperatures, changes are more rapid.

There is improved processing equipment and adequate refrigeration capacity in many wineries. White crushed grapes are held without SO_2 additions at lower temperatures with no adverse effects. Earlier experiments (Ough et al., 1969) had indicated that Chardonnay wines benefitted by some skin contact prior to fermentation. The optimum time was 12 hrs at 20°C but with SO_2 added. More recently, Long and Lindblom (1987) noted success without SO_2 additions and holding the grapes at 10°C for up to eight hours before draining and pressing. The color of these wines, without SO_2 added until after fermentations were complete, was as light or lighter than the companion wine protected from air and SO_2 added. The elimination of SO_2 is becoming quite common in clean white grapes before and during fermentation. With small lots, which are more prone to oxidation, this is questionable (Ough and Crowell, 1987). Nagel and Graber (1988) found no increase in color in wines from oxidized juice. If further excessive juice exposure to air occurred, losses in hydroxycinnamic and caffeic acids resulted. Oxidized wines were more bitter than wines made from unoxidized fruit.

Some winery experiments (Ramey et al., 1986) indicated that Chardonnay skin contact at 10°C resulted in wines superior to those held at higher temperatures. Some of the compounds noted, which underwent change, were esters, alcohols and phenolic compounds.

Baumes et al. (1988) also measured volatile components in white wine with skin contact. He found 32 compounds which increased 2.3 to 2.8 times over those in the free-run juice. All white wines may not benefit from skin contact but certainly with the temperature options now available, varieties with distinctive varietal characters probably do.

In normal winemaking, there is always some skin contact. In large bag or membrane presses, extended times occur as the juice is slowly separated. Some wineries using drainers have contact time in the ranges of hours. Probably more attention should be paid to the times and temperatures at which this contact occurs. For every 10°C rise in temperature, chemical rates of reactions approximately double. Biological reactions behave similarly.

Dubourdieu et al. (1986) show phenol increases in gallic acid, p-hydroxycinnamic acid and both cis- and trans-caffeic acids. They show amino acids after 18 hours of skin contact increased about 1.5 times. Proteins and terpenes also increased measurably.

More recently, Baumes et al. (1989) measured many of the critical wine aroma compounds in a fortified muscat wine. Among these were the terpenols, alcohols, esters and carbonyls including β-damascenone. They tested the effect of skin-juice contact before fermentation. Several hours at 25°C gave increased terpenols, esters and carbonyls with decreased amounts of fusel oils and other alcohols. The quality increases were confirmed by sensory evaluation. Phenolic increases were minimal and did not diminish the sensory appreciation of the wine. Similar results would be expected for dry wines. Ough (1969) showed changes in several compounds in some white variety wines as shown in Table 2-4. Intermediate times showed expected differences.

Table 2-4. Average Values[a] for Some Chemical Components and Fermentation Times for White Wines as Affected by Time on Skins.

Components or effect	Free run	Time on the skins (hours) 12	24	48
Fermentation time (hrs to 0°Brix)	132.0	119.0	119.0	114.0
Total acid (g H_2Ta/100 mL)	0.623	0.595	0.548	0.556
pH	3.28	3.42	3.55	3.62
Acetic acid (g/100 mL)	0.027	0.028	0.031	0.033
Tartaric acid (g/100 mL)	0.180	0.140	0.135	0.117
Potassium (mg/L)	763.0	898.0	1135.0	1174.0
Extract (g/100 g)	2.49	2.53	2.58	2.73
Ethyl alcohol (v/v)	12.69	12.72	12.70	12.49
Isobutyl alcohol (mg/L)	37.0	33.0	41.0	38.0
Isoamyl alcohols (mg/L)	282.0	281.0	271.0	251.0
2,3-Butanediols (mg/L)	579.0	668.0	678.0	792.0
Total vol. esters (mg/L)	48.0	47.0	48.0	51.0
Acetaldehyde (mg/L)	30.0	21.0	26.0	19.0
Total phenolics (mg/L)	317.0	371.0	420.0	571.0
Total nitrogen (mg/L)	177.0	185.0	213.0	217.0
Proline (as N mg/L)	81.0	93.0	98.0	106.0
Color (OD 420 nm)	0.133	0.171	0.204	0.241

[a]Average of nine different musts, each fermented under every condition. From Ough (1969).

pH REGULATION

There is very good reason to maintain a pH in fermenting wine below 3.5. The wines ferment more evenly, malolactic fermentation during the yeast fermentation is less likely and the wine has better sensory qualities. Above this pH, the effect of SO_2 in inhibiting wild yeast and bacteria is much reduced. In particular, this is important with red wine. Skin contact causes decreases in total acidity by the precipitation of potassium acid tartrate (Dubourdieu et al., 1986). The pH is buffered upward by the material from the skin. The addition of about 1 g/L of tartaric acid will lower pH about 0.1 unit. Trials should be run on samples before treatments. Juices from wines high in malic acid and high in potassium can have higher pH values as do wines with low tartaric and malic acids. Wine with low potassium values have lower pH values.

Iland (1984) applied Boulton's (1980) "extent of exchange" equation:

$$\text{Extent of Exchange} = \frac{[K^+] + [Na^+]}{[\text{Tartrate } (H_2Ta)] + [\text{Malate } (H_2Ma)]}$$

All the components are expressed as g/L protons $[H^+]$. Iland (1984) showed results from several Australian red juices from the application of this equation. Figure 2-8 shows the linear relationship of the "extent of exchange" to the pH.

SETTLING CONTACT TIME

Letting the juice settle longer before racking was found by Williams et al. (1978) to give more complexity to the wine. As long as SO_2 is added, this will inhibit the wild yeast (as long as the juice is cold) up to 24-48 hrs. Juice centrifuged and fermented quickly gave low-bodied fruity wines. Juices settled 24-48 hrs before racking resulted in more complex, higher bodied wines. The latter required longer maturation. Figure 2-9 shows a close-up view of the draining valve (lower). The racking valve (upper) allows separation of the clear juice from the settled lees.

USE OF SO_2

To protect the white grapes from oxidation changes, minimal amounts of sulfur dioxide (SO_2) can be added. This material is available in several forms, of which the most common, readily

FIGURE 2-8. The relationship between pH and the extent of exchange in grape juices. Sample points represent 17 'Shiraz' and 2 'Cabernet Sauvignon' grape juices (Data from Iland, 1984).

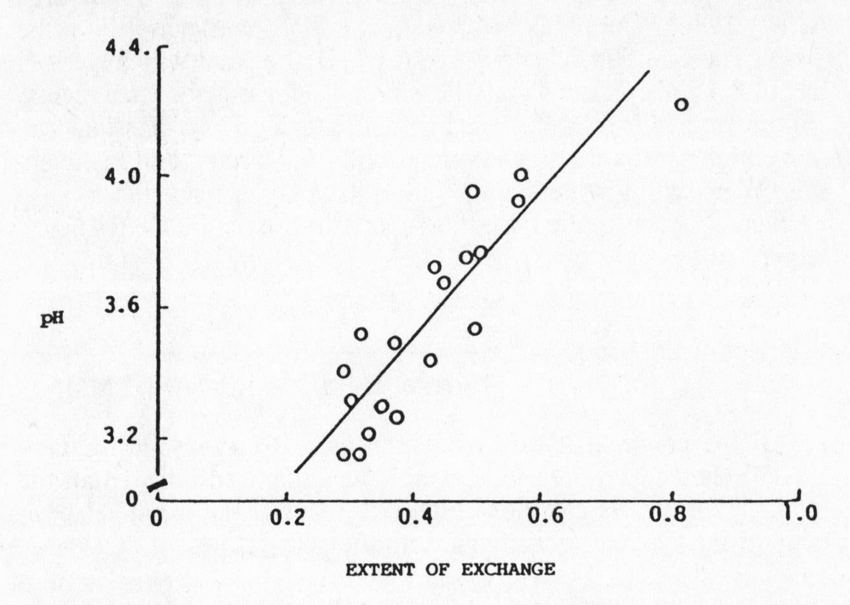

available and easy to use, are given in Table 2-5. There are other forms, but SO_2 gas and potassium metabisulfite are the best to use. The sodium form adds sodium to the wine which normally should be avoided. The gas form of SO_2 is available in cylinders. It is a very corrosive gas, all cylinder fittings should be of 316 stainless or equivalent noncorrosive metal. SO_2 gas is toxic and extreme care should be exercised when it is used. The preferred method of use for smaller operations is to saturate water with the gas. A 5% water solution is superior to using the gas directly from the cylinder. A plastic Tygon® tube from the regulator valve of the SO_2 cylinder to the bottom of a glass bottle containing water can be easily assembled. The gas is bubbled into the water at a moderate rate. When the bubbles reach the surface, the water will be close to saturation. With specific gravity hydrometer and a thermometer, check the solution. Compare the two readings to the data in Table 2-6 to calculate percent of SO_2 in the solution.

Adjust the concentration to a 5% ± 0.1% SO_2, either bubble in

FIGURE 2-9. Racking (upper) and draining (lower) valves of a tank used for draining off the clear juice from the settled pulp or lees.

more SO_2 or add more water. This solution is now ready to add to either the juice or the wine. For use amounts see Appendix I.

Using the metabisulfite is also simple, but requires that the amounts used be weighed out. When small lots are made, finding the proper scales for accurately weighing may be more inconvenient than using the 5% water solution.

Using the gas cylinder to put the SO_2 directly into the tank is simple. The tank must be such that the weight loss from the cylinder can be accurately measured. The addition of 100 mg/L for a thousand gals of wine is 0.834 lbs. Accuracy will depend on the weight of the cylinder and the sensitivity of the scale.

A winemaker has the option of either adding or not adding SO_2 at the crusher. The decision should be based on several factors. Are the grapes free of mold and rot? Are the lots of sufficient size so no excessive oxidation occurs because of large surface to volume ratios? If the answers to these questions are yes, then SO_2 addition can be eliminated, although the juice should be settled at less than 12°C.

Table 2-5. Chemical Forms of Sulfur Dioxide Used for Wine.

Name	Formula	% of SO_2
Sulfur dioxide (gas)	SO_2	100
Potassium metabisulfite	$K_2S_2O_5$	55
Sodium metasulfite	$Na_2S_2O_5$	65

Table 2-6. SO_2 - Water Specific Gravities Expected by Temperature for Three Concentrations.

Temperature °C	4 g/100 ml	5 g/100 ml	6 g/100 ml
15	1.020	1.025	1.030
20	1.018	1.023	1.028
30	1.014	1.019	1.024

It will ferment adequately if the proper amount of yeast is added and the temperature held to the desired level. If the answer is no, then some SO_2 (35-50 mg/L or more, if extremely heavy in rot and mold) should be added at the crusher.

In adding SO_2, one should be able to calculate accurately the amount. Figuring 200 gals of crush/ton of grapes and an amount of 37.5 mg/L of SO_2 added potassium metabisulfite (50% SO_2), it works out to be 1/8 lb/ton or 2 oz or 56.7 grams. The amount of SO_2 on the basis of dry gas SO_2 as 10 mg/L total of SO_2/1000 gals of juice or wine is 0.083 lbs/1000 gal. This can be measured by weight loss from the gas cylinder again if the tank being gassed is of sufficient size. Most larger wineries use proportional pumps to inject the SO_2 gas into the must line. These pumps are set to respond to the load on the must pump motor. Dosage equipment for adding SO_2 during juice or wine transfer is shown in Figure 2-10. If added properly and evenly during the transfer, further mixing is not required.

JUICE SEPARATION (HOME WINEMAKER)

When working with small amounts (less than 200 gals), it is easy to make mistakes. A certain amount of leeway exists. Whether the amount added is 60 mg/L or 85 mg/L of SO_2 makes little difference. However, if SO_2 is used, sufficient amounts should be added to inhibit the oxidizing enzymes. Usually 35-50 mg/L is enough for this, as well as for inhibiting the growth of wild yeast or bacteria present. Adding over 100 mg/L is unnecessary and may result in needlessly high SO_2 values later in the winemaking process.

Separation of the juice of the white grapes from their skins can be accomplished in several ways by the home winemaker. For a small operation, the basket press (Figure 2-11) or a small Willmes® press can be used effectively. Although there are better presses for commercial use, either of these is adequate for small operations. The presses must be operated properly for good results. With the basket press, some type of basket is necessary to hold the freshly crushed grapes. The crushed grapes are very slippery due to the relatively high pectin content. The basket press is effective if used in the following manner. Lay a layer of cheesecloth in the basket. Fill the

FIGURE 2-10. SO_2 dosage unit for adding SO_2 during wine transfer or directly into tank.

FIGURE 2-11. Small hydraulic basket press.

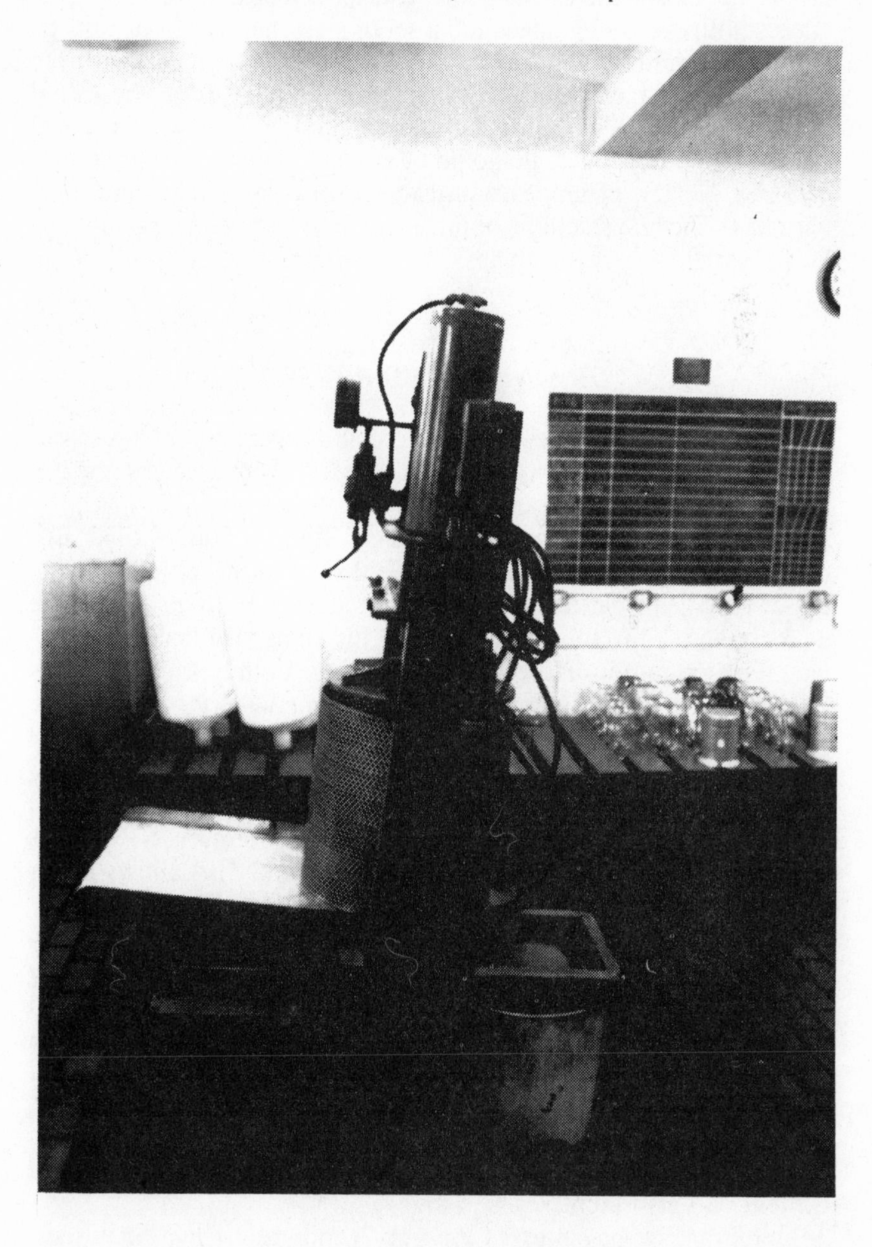

basket about six inches with the crushed grapes. Fold a layer of cheesecloth over the grapes, add a set of slats that will fit just inside the basket. Add another layer of cheesecloth. Continue to fill the basket in this manner. The first fraction of juice from the initial pressing should be set aside and the remainder of the juice pressed into a second container. It should be possible to obtain at least 140-160 gals per ton of grapes with careful pressing when using most varieties. The two fractions of juice should be fermented separately. Some varieties give much less juice on pressing. Sylvaner is an example.

DRAINERS AND PRESSES

There are several types of commercial drainers and presses. Because of the labor intensity of basket presses, there are very few left in commercial use.

The drainers most in use are Bucher® style or Potter®. The Bucher type has screens around the vertical sides of the tank, while the Potter type has a vertical center circular screen. Both do a good job of draining off the free-run juice. For the same size tanks, the Bucher type has more surface area for draining. Both types allow skins to slide out the slant bottom. Some tablewine wineries allow white grapes to be crushed in regular closed tanks, let the cap rise, and drain out the juice. The grapes are then removed for further pressing in several conventional type presses. Figure 2-12 shows a typical Australian drainer setup using Potter® drainers. Figure 2-13 shows a setup in California using refrigerated drainers. This allows better control of the quality of the juice.

Most of the presses have methods of separation of the juice by degree of severity of the pressing. The hard-pressed juice is very high in phenolic compounds as well as many other components that are expressed from the cells of the grape skins and pulp. These compounds have a very significant influence on the composition of the juice. The juice becomes bitter and harsh, the pH increases as well as the titratable acidity and the nitrogen content. If the grapes have significant amounts of raisins or shrivelled berries, the sugar content will also increase.

Membrane or bag presses are very gentle and leave little sedi-

FIGURE 2-12. A series of Potter® drainers for juice separation. Commonly used in Barossa Valley, Australia.

ment in the juice. An example is shown in Figure 2-14. Large Willmes® bladder presses (Figure 2-15) are also frequently used. The amount of solids expressed with the juice is slightly greater, but the process is more rapid.

Maillard reactions that can cause browning are controlled by temperature, pH and type of sugar. Very low and very high moisture contents decrease the formation rates drastically. Optimum rates are at/or about 15% moisture. Only reducing sugar can react. There is relatively little expectation of these reactions going on in wine. However, in pressing with high pressure presses such as screw

FIGURE 2-13. A set of drainers that are refrigerated. Chateau St. Jean in Kenwood, California.

FIGURE 2-14. A Bucher® membrane (or bag) press.

presses, it is not inconceivable that a Maillard reaction might proceed to some degree, although no reports have been published.

The hard-pressed juice is very high in solids and usually very difficult to handle. Foster and Cox (1984) reported on the successful use of a centrifuge decanter. This device continuously centrifuges out the solid material. Comparisons of the quality of the wines made from juice so handled were favorable. It lowered the phenol by an amount equal to a treatment of one lb/1000 gal of PVPP. In addition, there were cost savings on wear of the equipment from sand removed by a cyclone separator. The iron content of the product was reduced. The wild yeast levels were lowered to a point that pure inoculum added was effective. Overall losses in juice for this winery were cut by 10%.

The amount of juice yield depends on the cultivar and the condition of the grapes. For white grapes, the volume of free-run is about 120-170 gals/ton and the pressed fraction 15-60 gals/ton. Free-run will have 6.5% solids and this can be reduced to around 2% by

FIGURE 2-15. A Willmes® bladder press.

centrifugation or settling and racking. Press juice will have around 12% solids or more and this can be reduced to about 4% by centrifuging.

The lighter pressed juice does not always make the best wine. It depends on the "style" of wine that is desired. Some of the material expressed by moderate pressing can add body and a desired amount of flavor to the wine. Some additional phenolic components can contribute to a more desirable color to a white wine. The winemaker must have the flexibility of the different pressings. Later he/she can make blends to obtain the desired style of wine.

OXIDATION

The major oxidation substrates for grape juice are caffeol tartaric acid (caftaric acid), p-courmaroyl tartaric acid and feruloyl tartaric acid (Singleton et al., 1984). These are the browning precursors for wines made with a minimum of skin contact. Their report showed that caftaric acid was very easily oxidized to a new product while the other two tartaric acid esters were not changed as easily. Salgues et al. (1986) identified the oxidation product as S-glutathionyl-2,caffeoltartaric acid (Figure 2-16).

As long as excess glutathione is present, this product will not be oxidized to the o-quinone. It will be further oxidized to the 2,5-S-glutathionyl caffeoltartaric acid.

If the juice is oxidized without the protection of SO_2 to inhibit the polyphenol oxidase enzymes, these precursors and the resultant brown polymers from o-quinone reactions will precipitate. This stabilizes the browning reactions in the wines from this source.

PECTIC ENZYMES

To settle the juice more rapidly and to obtain a clearer product, add pectic enzymes at the crusher or as soon as possible. They will not only aid in the clarification of the juice, but also allow for easier pressing. It is not unusual to obtain up to 15% more free-run juice when the pectic enzyme-treated juice is held up to 12 hours after crushing before pressing (Figure 2-17). These enzymes break up the long chain pectin molecules into shorter more soluble compounds

FIGURE 2-16

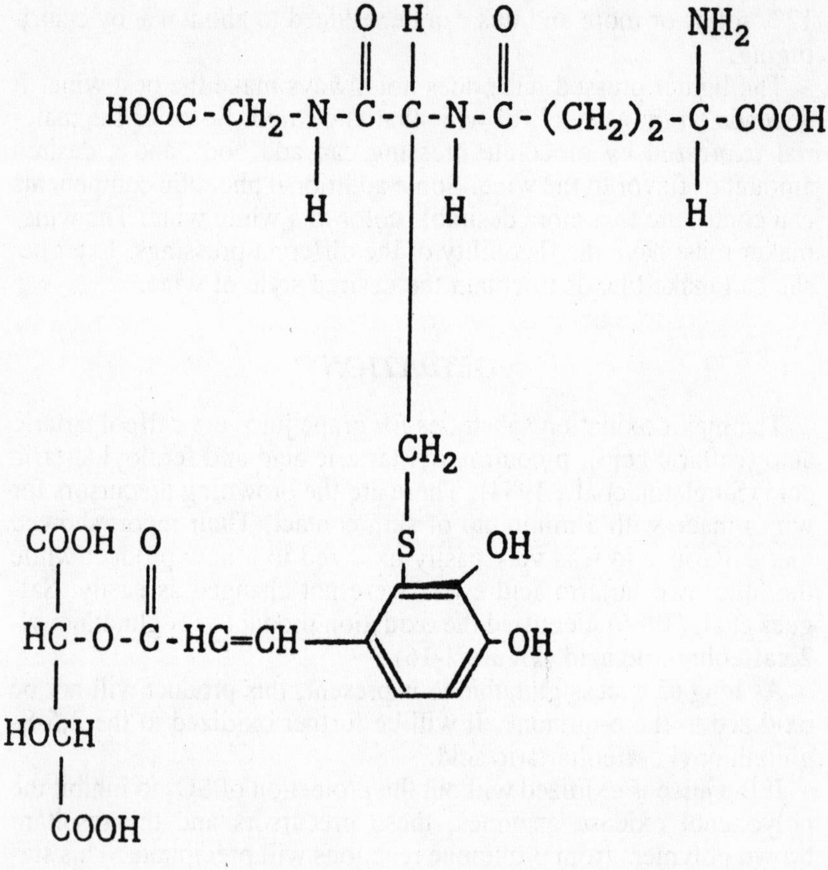

O H O NH$_2$

HOOC-CH$_2$-N-C-C-N-C-(CH$_2$)$_2$-C-COOH

H H H

CH$_2$

COOH O S OH

HC-O-C-HC=CH OH

HOCH

COOH

2-S-glutathionyl-caftaric acid.

which decreases the viscosity of the juice as well as clarifying it. The composition of the juice is not significantly changed by the enzymes except for the pectins and the methanol. Table 2-7 gives the changes in methanol with time and temperature compared to the control for several varieties.

Once the juice is separated from the skins, it is settled overnight

FIGURE 2-17. Juice-yield comparisons of control ◦ vs pectic enzyme-treated • lots at several skin-contact times for Chenin blanc ——— and for Muscat of Alexandria ------- (From Ough and Crowell, 1979).

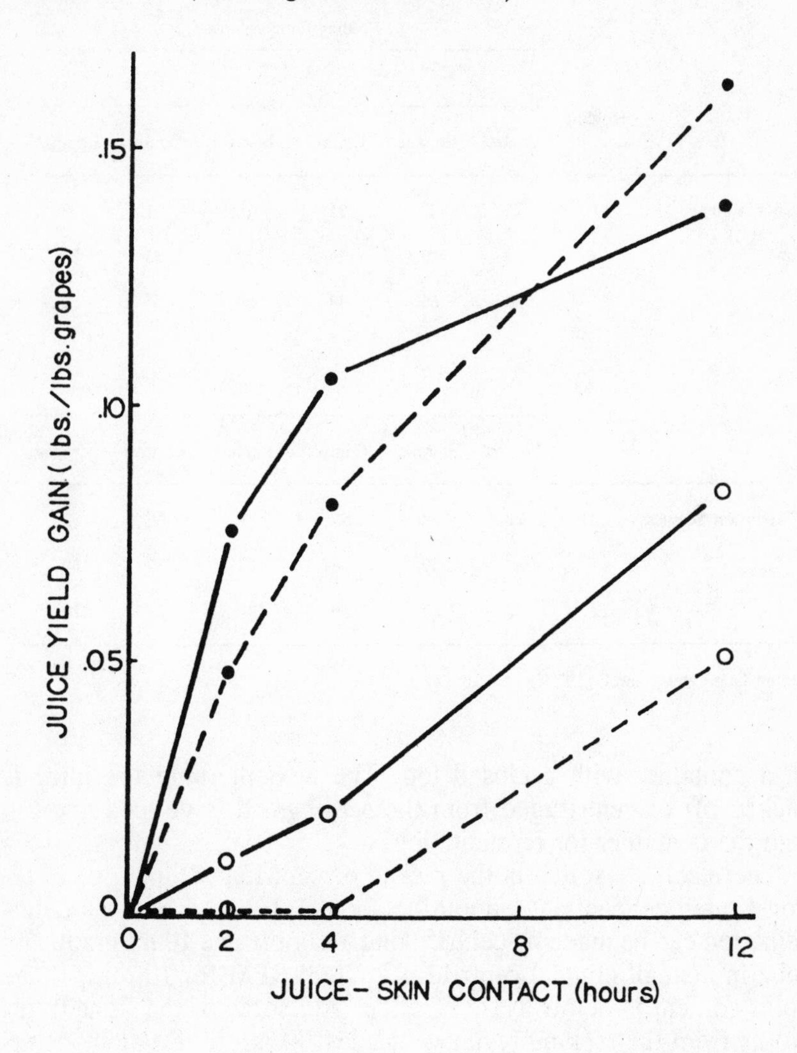

Table 2-7. Methanol Content mg/L as Affected by Time and Temperature with and without Pectic Enzymes Added.

		Holding temperatures					
		14.4°C		18.5°C		35.0°C	
	Holding time hrs	Control	Enzyme	Control	Enzyme	Control	Enzyme
Chenin blanc	0	25	27	21	18	31	39
	4	28	53	26	70	63	72
	12	30	66	34	64	58	87
		10°C		21.3°C		32.2°C	
		Control	Enzyme	Control	Enzyme	Control	Enzyme
Thompson Seedless	0	22	32	27	59	38	52
	4	30	59	46	78	50	101
	12	36	67	59	131	73	120

From Ough and Crowell (1979)

in a container with a closed top. The next morning the juice is racked off or centrifuged from the settlings. It is yeasted and put into the container for fermentation.

The insoluble solids in the racked or centrifuged juice, or in reconstituted concentrate, should be about 1 1/2% - 2%. This determination can be made by centrifuging a sample in a 10 ml graduated tube in a small clinical centrifuge at 2000 RPM for 10 min. If the solid content is below 1 1/2%, either add some of the light fluffy solids from the racking residue, or add 100 mg/L of diatomaceous earth or bentonite. The purpose of these solids is to give the yeast a surface on which to adhere, to enable them to float up and down as they ferment. This allows for a more uniform fermentation, giving more chance of the fermentation going to dryness.

JUICE ADDITIONS AND CORRECTIONS

One should decide if any amelioration is necessary. The total titratable acidity and the pH from the grape analysis should be considered. If the pH from the grapes is above 3.35-3.40, sufficient tartaric acid should be added to reduce the pH to between 3.25-3.35. This can be done by trials, taking a portion of the juice and adjusting the total acid upwards by 2 g/L, another by 3 g/L and another by 4 g/L. From these data you can estimate the amount of tartaric acid required to cause the desired change in pH. Tartaric acid added to the juice causes the reaction with potassium to form the insoluble potassium acid tartrate with the release of a hydrogen ion. The result is a lowering of the pH.

$$H_2TA + K^+ \longrightarrow KHTA + H^+$$

In addition, the H_2TA is in equilibrium with its undisassociated forms.

$$H_2 TA \rightleftarrows H^+ + HTA^- \rightleftarrows TA^{-2} + H^+$$

The final distribution of the forms depends on the final pH and the pKA of the tartaric acid. If no malolactic activity is anticipated, then malic or even citric can be used. Fumaric acid should not be used as it is metabolized by the yeast (Wagener et al., 1971).

If a pH meter is not available, there are pH papers with indicators impregnated in them, although this method is not as satisfactory. If there is a measure of the total acidity only, and not the pH, raise the total acidity to 9 g/L. This will probably change the pH sufficiently.

Red grapes also should be acidified before fermentation, if necessary. The basis for this is much the same as for white grapes. Addition of tartaric acid to raise the total acidity to 9 g/L (as tartaric) is usually sufficient. If malolactic fermentation is desired later, then the pH of the must, after adjustment, should not be below 3.3-3.4. Where the *vinifera* grapes do not ripen due to a short growing season, or if the grape were picked too early, it may be desirable to add sugar. The sugar should be added only in the amount necessary to bring the °Brix up to 23.5° maximum. The amount of sugar added will depend on the variety and the style of wine desired. Regular sucrose (table sugar) can be used. The yeast has the necessary enzymes to convert sucrose to glucose and fructose, the normal sugars in the grape.

If there are only French hybrids, labrusca or muscadine type grapes available, usually both sugar and water are added. The water is used to reduce the acidity and dilute the rather strong flavors associated with most of these varieties. These varieties seldom naturally reach the sugar content to make wines of acceptable alcohol levels for table wine.

NOBLE ROT

Under certain favorable conditions, the grapes in the vineyard can become infected with *Botrytis cinerea* "noble rot." The fruit can attain sugar values of 30-40°Brix. White Riesling grapes are the most susceptible to *Botrytis* infections. In most areas, it is rare for the conditions to be present that will produce *Botrytis* and then later dehydrate the fruit. The necessary conditions for *Botrytis* growth are rain/moisture on the ripe fruit. This is followed by a drying but cool environment. The *Botrytis* in the cool, dry cluster continues to grow. It draws the moisture necessary for its growth from the grape berries. This entire process can happen in less than a week when these conditions prevail. The berries should all be intact and shrivelled with no evidence of any other rot or mold. Other mold infections should be cut away and removed to produce a good Sauterne or Auslese type wine. Once the *Botrytis* clusters have been selected and trimmed, they should be pressed. The juice is treated with 100 mg/L of SO_2, settled and racked before yeast inoculation.

GRAPE CONCENTRATE (HOME WINEMAKER)

If it is clean and clear, grape concentrate can be fermented without any treatment other than pH and SO_2 adjustment and dilution. If the concentrate is cloudy after diluting, it should be settled and racked or centrifuged. Most concentrate is about 68-72°B, and needs diluting with "good" water about 5:1 to bring it to a desirable °Brix.

GRAPE JUICE COMPOSITION

Sugars

The sugars in grape juice, except for the water, make up the predominant constituents. The major contributors are glucose and fructose, fairly evenly distributed when the grapes are mature. Minor amounts of galactose, mannose, cellobiose, melibiose, raffinose, arabinose, rhamnose, xylose, ribose and fucose have been found. Sucrose is seldom found in the juice. Sucrose is hydrolysed at the pH of the juice and wine. It is enzymatically split by the plant during its transport to the berry from the leaf. Total reducing sugar in the grapes usually will not exceed about 250 g/L. In any juices with higher amounts than this, the extra sugar is probably due to shrivelling of the berries. This can be caused by water stress in the vine, dehydration by excess heat or *Botrytis* mold activity. The latter can cause the sugar concentration to increase to over 40°Brix. The other reducing sugars of mention are L-arabinose and D-xylose. These can range from about 0.2-1.5 and 0.03-0.1 g/L, respectively. The polysaccharides found in grapes that cause colloids have been reported by Kishkovskii and Skurikhin (1976) as pentosans 0.03-2.0, pectins 0.5-2.0, gums 0.3-4.0 and dextrin 0-1.1 g/L.

Acids

The organic acids found in grapes represent the next largest group of compounds. During the period near sugar maturity, the tartaric acid is the predominant acid in the warmer climates (2.0-8.0 g/L). This is because the malic acid respires significantly at 40.5°C. Tartaric acid respires at 57.2°C or higher. In the cooler climates, in grapes picked at early maturity, the amounts of malic acid can exceed those of tartaric acid. On the average, malic acid at grape maturity accounts for 10-40% of the acid fraction. In early season, it may constitute as much as 60% of the acid fraction. Small amounts of citric acid are present in the juice (from 0.1-0.7 g/L. Other acids of the citric acid cycle are also present in just trace amounts.

Ascorbic acid is present from 5-15 mg/L in grapes. If one looks hard enough, many organic acids can be identified, but they are of minor consequence.

Acids formed by action of molds on fruit can be of significance. Gluconic acid in normal juices is less than 0.5 g/L; while in botrytized grapes, up to 5.9 g/L. Must can contain up to 4.0 g/L of uronic acids (95% galacturonic); but in botrytized grapes, up to 7.5 g/L.

Sugar Alcohols

The sugar alcohols can be found in significant amounts both in table wines and wines infected with *Botrytis*. The values reported for sorbitol are 5-194 mg/L in table wines and 84-989 mg/L for botrytized fruit. Mannitol can come from lactic acid bacteria spoilage with the reduction and dephosphorylation and dehydration of fructose-1-PO_4 or from the activity of *Botrytis* mold. After only two days of *Botrytis* infection in a sample, the mannitol went from 82 to 536 mg/L. Meso-inositol is present in grapes around 500 mg/L. Other sugar alcohols present at 25-50 mg/L are erythritol, arabitol and trehalose. *Botrytis*-infection increased all these components. Storage of juice from *Botrytis* infected fruit caused further increases in these components according to Dubernet (1974).

Nitrogen Compounds

Nitrogen compounds are mainly amino acids with smaller amounts of protein, peptides, ammonia, nitrate, amines and amides with smaller amounts related to vitamins, cofactors, pyrimidines, purines, nucleic acids, glycoproteins, hexosamine, etc. Protein can vary from <20 mg/L to >100 mg/L, and this component is an important factor for wine stability. There has been relatively little data reported on the peptides in juices, but they are present and could be important sensory components. Ammonia in the grape juice can vary from 20 mg/L to >300 mg/L with the average values around 100 mg/L from healthy vines. Amounts can vary with cultivar (Ough and Kriel, 1985). Nitrate in the juice varies from <1 mg/L up to over 100 mg/L with an average of about 10 mg/L. Ranges and averages for amino acids in commercial grape juices are given in Table 2-8. It is noticeable that the major differences are in the proline/arginine ratios. White juices are from pressed grapes while the

Table 2-8. Amino Acids and Ammonia (as mg/L) Values on Juice from the Napa Valley and Carneros Wineries. 1988 Season.

	Sauvignon blanc		Chardonnay		Pinot noir		Cabernet Sauvignon		Zinfandel		Merlot	
Amino acid	Range	Mean	Range	Mean	Range	Mean	Range	Mean	Range	Mean	Range	Mean
Aspartic acid	15-112	42.6	10-106	40.8	3-122	38.2	5-51	17.5	7-86	24.4	1-29	18.6
Threonine	28-119	67.1	28-152	73.0	19-249	102.3	0-80	32.0	13-76	33.1	4-62	30.9
Serine	31-105	65.8	26-252	98.7	11-150	67.9	5-89	37.5	13-53	32.1	7-74	45.8
Asparagine + glutamic acid	43-227	130.0	27-282	123.6	13-345	138.9	12-102	42.3	40-156	70.0	7-68	34.1
Glutamine	40-744	207.4	38-1336	248.1	43-651	196.8	12-261	57.0	7-259	90.3	5-138	55.7
Proline	83-2302	431.0	145-2674	911.2	55-500	280.1	780-2973	1775.3	306-2286	856.1	880-2415	1689.7
Alanine	34-300	160.3	33-514	207.6	70-324	179.7	14-143	54.7	18-89	51.1	19-84	47.9
Citrulline	0-48	13.0	0-18	2.9	0-49	14.7	0-30	3.3	0-23	5.9	0-18	6.9
Valine	14-81	40.3	15-98	38.7	6-79	37.7	2-78	28.9	8-43	20.2	2-52	26.6
Methionine	1-28	8.4	3-32	10.1	0-32	11.5	0-29	6.6	1-9	4.1	0-25	6.1
Isoleucine	7-62	19.9	3-67	17.4	0-59	21.1	1-45	14.9	6-27	12.8	1-30	14.4
Leucine	6-73	25.8	0-72	22.7	0-69	26.9	0-68	18.2	0-30	13.5	0-38	18.6
Tyrosine	9-46	23.4	9-68	29.7	0-27	15.3	1-34	12.7	8-25	12.6	5-22	11.2
Phenylalanine	10-102	35.4	7-122	28.7	0-48	20.5	1-46	14.0	6-26	13.5	1-21	10.4
Ammonia	36-188	94.1	46-180	106.5	54-249	119.0	21-226	82.8	50-153	89.7	14-115	69.2
Histadine	10-75	38.5	9-126	45.5	2-58	28.7	7-43	20.7	9-41	23.3	2-36	19.7
Ornithine	1-34	10.0	0-45	4.3	3-44	14.9	0-20	4.4	0-14	5.1	2-25	11.3
Lysine	1-28	11.5	0-26	9.6	0-28	9.3	0-25	8.6	0-26	8.3	0-22	9.3
Arginine	210-1432	889.9	186-1377	513.6	280-2808	972.5	63-707	198.7	73-919	483.1	65-1282	406.2
Total	632-4501	2314.4	1012-6347	2532.0	795-5371	2296.0	1016-4693	2430.1	936-3564	1799.2	1276-3514	2532.6

The number of juice samples from 22 wineries were 39, 94, 50, 122, 28 and 19 for Sauvignon blanc, Chardonnay, Pinot noir, Cabernet Sauvignon, Zinfandel and Merlot, respectively.

1) Total does not include γ-amino butyric acid

2) Zinfandels were on the whole picked early for blush wines which probably accounts for the lower total and relatively low proline. Both increase with fruit maturity.

reds are from the free-run juices. This may partially account for differences in amounts of some of the amino acids as skin contact or pressing can double the amino acids in some cases.

Millery et al. (1986) measured the amino acids of must of Chardonnay, Pinot noir and Pinot Meunier in the Champagne area. They could separate the Chardonnay statistically from the Pinot noir and Pinot Meunier. The latter two were more similar in amino acid makeup. Their data for Chardonnay and Pinot noir are summarized in Table 2-9 for comparison with the data presented in Table 2-8 for Napa Valley. Part of the fruit from Napa Valley was harvested more mature and was for table wine, not champagne. Some differences do appear. The proline is greater for the California musts as is the

Table 2-9. Amino Acids (mg/L) from Musts of Chardonnay and Pinot noir.

	Chardonnay[2]		Pinot noir[3]		
Amino acid	1983	1984	1983	1984	1987[4]
Aspartic acid	25.1	24.7	36.8	32.8	134.2
Threonine	90.3	147.4	131.1	136.7	178.4
Serine	157.2	211.4	88.8	134.6	125.8
Glutamic acid	66.3	86.9	74.6	73.2	406.7
Glutamine	410.9	973.2	218.8	626.0	-
Proline	611.2	329.4	119.8	109.9	232.8
Alanine	366.2	527.2	264.2	424.0	89.9
Citrulline	18.7	37.4	36.3	94.9	-
Valine	43.0	119.0	38.5	73.0	77.4
Cystine	10.8	2.4	1.1	0.0	0.0
Methionine	5.2	44.1	5.4	24.3	9.2
Isoleucine	24.2	124.5	35.0	90.2	64.3
Leucine	23.7	139.3	36.3	106.8	103.9
Tyrosine	12.1	27.9	10.1	18.5	44.9
Phenylalanine	31.8	126.9	35.5	84.7	78.7
Ornithine	0.0	3.3	6.5	14.6	9.7
Lysine	10.3	7.9	13.7	9.9	6.5
Histidine	26.4	42.7	23.1	33.5	42.5
Arginine	220.5	374.8	574.4	695.3	737.0
Total	2258.0[1]	3414.0[1]	1859.0[1]	2873.0[1]	2629.6

[1]Includes ethanolamine, γ-aminobutyric acid, β-alanine but not ammonia.
[2]For the Chardonnay 10 samples in 1983 and 65 in 1984.
[3]For the Pinot noir 9 samples in 1983 and 27 in 1984.
[4]Pinot noir grape must from Bourgogne, 15 samples, Ooghe and Kastelijn (1988). Includes α-amino butyric acid 23.9 mg/L and glycine 71.2 mg/L.

arginine. The alanine and glutamine are considerably less. There are some differences in citrulline and leucine and isoleucine. Maturity differences may explain the differences, as may different rootstocks, fertilization regimes and other factors of climate, vine training, clones, etc. Table 2-9 also includes some results of Ooghe and Kastelijn (1988) for Pinot noir from the Burgundy region of France. Some discrepancies appear in the values compared to the other two sets of results.

The vitamins are present in minor amounts but sufficient for the yeast's needs (Ough et al., 1989). Biotin, the only one the yeast cannot synthesize, is present in abundance; also, the yeast accumulate significant amounts and store it for future generations. If pantothenic acid is scarce, the yeast can make some H_2S in the process of synthesizing it. Other nitrogen components are present in small amounts.

Phenolic Compounds

The phenol content of juices depends on how they are measured to some extent. If they are measured spectrophotometrically, the reports show around 100 mg/L, but by the Folin-Ciocalteau method as gallic acid equivalents (GAE), about 150-250 mg/L. The phenol present in white juices are catechins, simple non-flavonoid phenols, flavonols and flavonones with some flavan-3,4-diols. No matter how quickly and gently red juice is separated from the skins, some traces of anthocyanins and other phenols are present. The phenols present in the white juice are all that is present in the wine. If there is skin contact or barrel aging, then more phenols appear. With the red wines and rosés, significant amounts of phenols are picked up from the skins during fermentation.

Other Compounds

In the juice, only a few alcohols are present. The main one is hexanol at around 4 mg/L. However, once grapes are crushed, the natural esterases present in the grape start to demethylate the pectins. The amount of methanol that rapidly forms is 30 mg/L or more. With extended skin contact, the values can go well over 100 mg/L (Table 2-7).

Roufet et al. (1987) found in four cultivars (Table 2-10) that the level of fatty acids in the grapes was 0.045%. The main ones were unsaturated acids. Linoleic, palmitic, linolenic and oleic were most common, followed by stearic and behenic acids. Phospholipids formed the largest fraction with neutral and glycolysides next. Glycolipids had a high percentage of linolenic acid, while with the phospholipids and the neutral lipids, linoleic was most abundant. No differences were found between varieties. The lipids were 1.5 to 3 times more abundant in the skin than in the pulp.

Zironi et al. (1984) identified C_{21}-C_{33}, odd and even, in addition to the C_{12}-C_{20} acids. The total amounts were small. Also, C_{17} was noted and myrisloleic C_{14}:1. The sterols and triterpene-diols percentage of the fractions were calculated. Beta-sitosterol made up

Table 2-10. Acid Distributing mg/100 g Berries in the Lipids (Four Cultivars Averaged) over the Maturity Range.

	Neutral lipids			Glycolipids			Phospholipids		
Acids	Green	Ripe	Over-ripe	Green	Ripe	Over-ripe	Green	Ripe	Over-ripe
Palmitic	1.10	1.65	1.80	0.95	0.80	0.85	7.20	7.70	7.50
Steric	0.25	0.40	0.45	0.30	0.30	0.30	0.30	0.90	0.90
Anachidic	0.15	0.25	0.40	tr	0.05	0.10	0.20	0.35	0.25
Behenic	0.20	0.25	0.30	tr	0.05	0.15	0.30	0.45	0.45
Total saturated	1.70	2.25	2.95	1.25	1.20	1.40	8.50	9.40	9.10
Palmitoleic	0.15	0.30	0.20	0.10	0.05	tr	0.10	0.10	0.10
Oleic	0.90	1.80	1.75	0.30	0.50	0.80	1.50	4.00	4.00
Linoleic	4.70	3.50	3.50	1.40	1.10	1.50	13.50	14.20	15.10
Linolenic	2.40	1.60	1.60	5.10	2.70	1.80	4.00	3.80	3.90
Total unsaturated	4.15	9.75	10.00	8.15	5.55	5.50	27.60	31.50	32.20
Total acids	5.85	12.00	12.95	9.40	6.75	6.90	36.10	40.90	41.30

65-88%, compesterol 5-15% and stigmasterol 3-14%, with six others making up lesser amounts. The absolute values were not given.

The main inorganic cation found in juice is potassium. It can vary from <400 mg/L to over 2000 mg/L. The amount is dependent somewhat on the soil, irrigation conditions, cultivar and climate. The other major cations are Mg^{++}, Na^{+}, and Ca^{++}. The sodium is present at low levels, 10-20 mg/L, and the Mg^{++} and Ca^{++} at 30-150 mg/L ranges. Other cations, iron and copper can be fairly high, 5-10 mg/L or more in the juice. The yeast will adsorb most of them during fermentation. There are numerous other cations but most are measured in wines more often than in juices.

The main anions are sulfate (as K_2SO_4) at 1 g/L, phosphate around 300 mg/L and chloride at 70 mg/L. Other anions are there in very small amounts.

There are literally hundreds of other compounds, but these only make up a very small percentage of the total.

Chapter 3

Fermentation and Wine Composition

The best wine can only be made from the best grapes, but it is not difficult to make poor wine from good grapes. The fermentation process is one of the most important steps in winemaking. Care must be taken throughout the process of winemaking, but during this step major damage can occur.

YEAST

Yeasts are a large group of one-cell organisms. They can exist under a myriad of conditions and can utilize numerous substances as their food source. The small group of yeasts that we are interested in for winemaking is the genera *Saccharomyces*. This group is commonly used for the fermentation of the juice. Only certain species of *Saccharomyces* are used. The most common is *Saccharomyces cerevisiae*. There are two commonly used races of *S. cerevisiae* – race *cerevisiae* and race *bayanus*. There are many strains of these yeasts. One of the more commonly used is *Saccharomyces cerevisiae* race *cerevisiae* strain Montrachet. The strain name usually, but not always, designates the area from which the yeast was originally obtained or what it is used for. With the current mobility of winemakers and winery microbiologists, many of the same yeast strains may be given different names and their original identity lost. For example, the Montrachet yeast was obtained by the Department of Viticulture and Enology, University of California at Davis, from the Geisenheim Station in Germany in the 1940s. The microbiologist in Germany had obtained it from a vineyard in Montrachet. It was given the number 522 when added to the departmental yeast collection at Davis. While visiting an Australian winery that had a yeast which was called V-22, the history of this yeast was requested. The only facts known were that it was a California yeast and had the number 522 on the original test tube. It is very possible that different strains of yeast, used in various parts of the world, came from the same yeast selections but now carry entirely different designations. Strain differences are usually based on a less accurately defined characteristic than that of species and variety. Species, or variety, or races can be differentiated by shape, sugars, acids and alcohols they will ferment, and more recently by gel electrophoresis of the soluble proteins (van Vuuren and van der Meer, 1987). Strain differences are based on such things as rates of fermentation, the ability to ferment a juice to dryness, the aroma or flavors produced in the wine, and the degree to which hydrogen sulfide is produced. However, with the new methods being developed to determine the DNA fingerprinting, even small differences otherwise unnoticed can be determined. Even careful analysis of the yeast's long-chain fatty acids can separate some strains. This method can easily identify *Saccharomyces cerevisiae* from other yeasts (Tredoux et al., 1987).

Identification

Petering et al. (1988) reviewed the methods for identifying yeast strains of wine yeasts. He paid particular emphasis to DNA methods which are not influenced by cultural conditions. Using transverse alternating field electrophoresis, 10 of 12 yeasts strains would be positively identified by chromosome size polymorphism. The methods for doing this type of DNA fingerprinting involves splitting the DNA molecules into fragments by endonuclease enzymes. These fragments are separated into bands on agarose gel by the special electrophoretic treatment. This allows the smaller and large fragments to be moved apart but kept near to each other.

This type of work has led to a much reduced number of classified species of *Saccharomyces*. Recently *S. acetic, bayanus, capensis, chevalieri, corneanus, diastaticus, globosus, heterogenicus, prostoserdovii, steineri* and *uvarum* have all been reclassified to *S. cerevisiae* and *S. delbrueckii, fermentati* (flor sherry yeast) and *rosei* have been classified as *Torulaspora delbrueckii*.

For those who want an idea of how to store, culture and identify yeasts by classical methods, a good beginner's article is that by Campbell (1988). For those interested in more detail, there are good texts on the subject (such as Dittrich, 1987).

A list of the scientific names, related information and some comments about a few of the yeasts associated with wine are given in Table 3-1. This is not a complete list of yeasts, but includes the main yeasts that are important to enologists. Literally hundreds of species, races and strains have been isolated from juice and wine.

Choice and Propagation of Yeast

The choice of a yeast is important. Each can give slightly different nuances of aroma and flavor to the wine. Trials are never completed in searching for the perfect yeast. There is always a new yeast to try. Some yeasts have the ability to make more esters, others more fusel oils (higher alcohols). Other yeasts require less nutrients and some ferment faster than others. One important factor is foaming. This is somewhat dependent on the medium, but it also depends on the yeast. Every winemaker has his/her own choices for certain jobs. For example, there is a "distillers" yeast, a "flor" yeast, "champagne" yeast, etc. These yeasts were selected over the years because they were most suited for the purpose. With the use of dried yeasts, the selection has become smaller, but the choices are made somewhat as before. Commercial wineries use from one to five or six different yeasts. One yeast for certain white wines and another for red wines. Different cultivars of the same color are sometimes fermented with different yeasts.

Nagel et al. (1988) did a comparison of six commercial yeasts and determined Prise de Mousse was best overall for certain limited conditions.

Some yeasts can cause problems as juice contaminants. For example *Saccharomyces cerevisiae* race *uvarum* can naturally produce up to 90 mg/L of SO_2 according to Ciolfi (1988). For 25 strains of this yeast, the average SO_2 produced was 56.3 mg/L.

The home winemaker, as well as the commercial winery, can now obtain and inoculate with a dried yeast. This is more satisfactory than letting the natural yeasts take their course, as in the past,

Table 3-1. Yeast Commonly Found in or Used in Wine.

WINE YEAST	COMMENTS
Saccharomyces cerevisiae var. cerevisiae Examples of strains: Montrachet California Champagne Epernay II	The yeast form nearly round to oval cells with elongated buds in grape juice. Colonies on malt-agar media are usually creamy white.
Saccharomyces cerevisiae var. bayanus Examples of strains: Prise de Mousse Pasteur champagne	This yeast forms more rounded buds than var. cerevisiae. It can be used as a "flor" yeast and is used as a "champagne" yeast.
Zygosaccharomyces bailii	This yeast grows well as a contaminant and is tolerant to very high levels of sulfur dioxide. Resistant to 600 mg/L sorbic acid and to alcohol. It is osmophylic as well and will grow in concentrate. The cells are smaller than the previous two varieties but difficult to distinguish.
Torulaspora delbrueckii Flor yeast	This is a film yeast used for flor sherry. It grows well on the surface as well as a rapid fermenter at warmer temperatures. It grows slowly when taken from yeast culture slants.
Schizosaccharomyces pombe	This variety forms cylindarical cells with the division at the middle of the cell. It will use malic acid as a substrate. It is a very slow fermenter and produces "off" tastes.
SPOILAGE YEASTS	
Hansenula anomala	These small oxidative yeast can multiply rapidly in juice with insufficient sulfur dioxide. They are small cells which are occasionally dumb-bell shaped. They produce excessive amounts of ethyl acetate.
Pichia membranaefaciens	They form long thin cells which may form chains. They grow under oxidative conditions and are not tolerant to alcohol or sulfur dioxide.
Brettanomyces sp.	These yeast contaminate wines, especially red wines which have not had enough sulfur dioxide added. A distinctive "off" smell and taste results.
Candida vini	This yeast, or similar species, form on the surface of newly fermented wines if air is not properly excluded. The result is a rapidly oxidized wine.

or preparing cultures from slants. If proper facilities are available to propagate the yeasts, the winemaker can obtain slants of yeast cultures from supply houses or from wine shops which cater to these needs.

Dry yeast comes in packages similar to dry bread yeast and is prepared for use in the same manner. The dehydrated yeast must be properly rehydrated before inoculation of the juice. This is done by bringing water to about 40°C (104°F), adding the dried yeast, letting it dissolve and rehydrate for about one hour. The amount to use should be sufficient to give a viable cell count in the must of 1×10^6 to 5×10^6 cells/ml. Most dried yeasts have about 10^{10} viable cells/gram (1/28th of an ounce) if they are prepared properly. If you put one gram of dried yeast into 100 mL of water, you will have 10^8 cells/ml or 10^{10} cells/100 ml. So, if you have five liters of juice, you will use the whole 100 ml of the culture.

To calculate the number of grams of dried yeast needed for varying amounts of juice, use the following formula:

$$\frac{\text{ml of juice} \times 2 \times 10^6 \text{ cells/ml juice}}{10^{10} \text{ cells/gram of dried yeast}} = \text{grams of dry yeast needed}$$

For commercial purposes, the yeast comes in large containers and is rehydrated and used directly. A tank of juice previously treated with SO_2 can be used to build a starter culture. This juice should be clean and relatively free of wild yeast if it is used to build up a starter culture.

If the directions for rehydrating the yeast are not correctly followed, either a delayed fermentation or a wild yeast fermentation may occur. The yeast cannot tolerate temperatures over 40°C and will even be killed at that temperature if held for any extended length of time.

It is more complicated to culture your own yeasts from slants. Some knowledge of sterile techniques and microbiology is necessary. The equipment to sterilize the juice, glassware and culture media is required, as well as a microscope.

Slants of yeasts grown on malt agar are the source of the starting material. The whitish material that grows on the surface is the yeast. Grape juice diluted to 50% with water is sterilized. This can be done in a pressure cooker. About 100 ml of the diluted juice at/or about 10°Brix, is poured into a 250 ml Erlenmeyer flask and the top closed with a plug of nonabsorbent cotton. This is put into the pressure cooker (or autoclave) and held at 10 pounds of pressure for 15 minutes. Several flasks can be prepared at the same time. The flask is allowed to cool. A small portion of the sterilized juice is put into the sterile culture slant. The yeast is washed off into the sterile flask of juice. This should be repeated several times to remove most of the yeast from the culture slant. The cotton plug is put back in the flask. The flask is kept in a warm place (25-35°C) for several days until the culture is growing vigorously. The culture should have a cell count of about 10^8 cells/ml or a total of 10^{10} viable cells/100 ml. This culture is the amount necessary to inoculate up to five liter of juice to give 2×10^6 cells/ml. For commercial purposes, the culture would have to be expanded to fit the need. Such equipment is shown in Figure 3-1 and can accommodate the needs of a large winery.

Yeast cultures can be grown with or without the presence of oxygen. For the most favorable results, some exposure to air is desirable. Yeast cells will live longer and ferment better when grown in the presence of oxygen. The reason is the production of sterols, which are involved in cell health. These are formed during aerobic activity of the yeast.

It was further confirmed by Sablayrolles and Barre (1987) that oxygenation of a fermentation is best done after the exponential growth phase of the yeast has finished. They suggest a minimum of 10cc O_2/L. Faster fermentations were found if 20 or 30cc/L were added. This was with model solutions and would probably hold true if the must lacked sterols (which their medium did).

Yeast cultures grow exponentially. They double in number at set intervals until they reach 10^7-10^8 cells/ml when exponential growth ceases. The period of exponential growth is the best time to use the culture for an inocula. Dry yeasts are acclimatized to SO_2 and will show little delay if 35-75 mg/L of SO_2 is used. Yeasts propagated from slants should be transferred into juice containing 50-100 mg/L SO_2 to be sure a tolerant culture is achieved. Yeasts can suffer from shock. Sugar, temperature or alcohol can contribute. When this happens, the yeasts may spill out some of their nitrogen content and be delayed in multiplication.

Radler (1987) discussed the use of dried yeast cultures. Measuring viable cells in dried yeasts, he found the level varied from 4×10^9 to

FIGURE 3-1. A commercial winery yeast propagation system with a jacketed mixing tank and pumps for transfer and aeration.

4.6×10^{10} cells/g with *Leuconostoc* contamination from $<10^3$ to 8×10^6 cells/g. He also discussed killer yeasts. For a particular killer strain, three pg of toxin was required to kill a sensitive yeast cell. Each killer yeast cell produces only 1/100th enough toxin to kill a single sensitive cell. The normal inoculum is usually more than 10^6 cells/ml for a grape juice. For a wild killer strain to cause growth problems for a sensitive yeast inoculum, the killer yeast load in that juice would have to exceed 10^5 cells/ml. The glycoproteins that make up the toxin are less effective at lower pHs of wines and temperatures over 30°C. The molecular weight of these protein toxins are around 11,000 to 14,000 daltons and can be partially adsorbed by bentonite (Radler and Schmitt, 1987).

The killer yeast toxin is produced by a virus-like particle occurring in the yeast cytoplasm. Other reports have noted that the toxin can be from as little as 0.2% to 2.5% of the total yeast population to cause killing of sensitive strains (not all strains are sensitive). This causes stuck fermentation (van Vuuren and Wingfield, 1986 and Tredoux et al., 1986). Recent reports by Jacobs et al. (1988), showed that killer yeast strains are quite commonly found on grape skins. They also noted that K_2 and K_3 strains were the only ones effective on sensitive strains of *Saccharomyces cerevisiae* at the pH values of wine or grape juice. Methods were investigated for screening and identifying killer strains.

Killer yeasts are rarely a serious problem. They could be if methods of fermentation change and more continuous fermentations are attempted. Radler and Knoll (1988) found a common applicculate yeast, *Hansenula uvarum*, that produced a toxin. It caused a ten-day delay in a *S. cerevisiae* fermentation.

INHIBITION

Fungicides and other pesticides are capable of inhibiting the growth of the yeast after it is added to the juice. The most common inhibition is caused by SO_2. An error in addition, or a yeast which cannot tolerate this substance, is the main cause for failure of yeasts to grow in grape juice. If the juice is completely free of oxygen, then there can be delays in growth but under commercial conditions this seldom, if ever, happens.

Sulfur Dioxide

The SO_2 activity is mainly associated with the free SO_2. The amount of free SO_2 can be measured easily. This consists of the $SO_2 \cdot H_2O$, HSO_3^- and $SO_3^=$. Of these, the $SO_2 \cdot H_2O$ is the active form. By use of the modified Henderson-Hasselbalch equation, good estimates of the $SO_2 \cdot H_2O$ can be made. This is commonly called the molecular SO_2.

$$SO_2 \text{ mg/L} = \frac{\text{Free } SO_2 \text{ mg/L}}{[1 + 10^{(pH-pKa)}]}$$

Depending on the yeast's tolerance, the amount of free SO_2 required to inhibit its growth at any given pH can be calculated. The pKa for SO_2 is 1.76. For *Saccharomyces cerevisiae* SO_2-tolerant yeasts, the levels can be significant as well as for *Zygosaccharomyces bailii*. For wines, the amount of $SO_2 \cdot H_2O$ required for *S. cerevisiae* has been given from 0.8 to 1.5 mg/L as SO_2.

Sudraud and Chauvet (1985) reinvestigated the ability of molecular SO_2 to prevent yeast growth in some commercial wine cellars. They determined a molecular value of 1.5 mg/L to stop fermentations. It took 1.2 mg/L in aging wines to prevent regrowth.

Loiseau et al. (1987) genetically marked the yeast inoculum so they could follow the growth during fermentation. They compared it to the wild yeast population growth. In an experiment to determine the effect of SO_2 addition at inoculation, or after 18 hours (18-20°C), they found each was equally effective. In studying normal thermovinification and maceration carbonique, only maceration carbonique had problems with contamination. This occurred later in fermentation and was 20-30% contaminated by non-inoculated yeasts. For the inoculated yeast to outgrow the indigenous population, normal inoculated must without SO_2, they found the ratio of the added yeast/indigenous yeast needed to be greater than 12/1. It was safer at 300/1 ratio.

Fungicides

It is a common misconception that fungicides used on the grapes can cause stuck fermentations. Most of the fungicides either kill the yeast inoculum, or inhibits its initial growth. Once the yeasts overcome the initial inhibition, they usually grow satisfactorily and ferment normally. There may be exceptions when the level is quite close to the toxic level. Lengthy delays in yeast growth can occur because of partial killing of the yeast inocula.

One of the first fungicides to be determined as a yeast inhibitor in grape juice was Captan®. Since then, a number have been investigated. Table 3-2 lists some that inhibited two common wine strains of *S. cerevisiae*, Montrachet and Steinberg. The test was for 16 hours. Yeast can grow out of short-term inhibition and ferment quite normally. Chiba et al. (1987) noted the order of decreasing inhibition of several fungicides on yeast growth to be Captan® >

Table 3-2. Some Fungicides and Pesticides which Inhibit Wine Yeast.

Trade name	Active ingredient	Level at which inhibited
Dithane M-22®	80% maneb	1-10 mg/L
Dithane M-45®	80% mancozeb	1-10 mg/L
Karathane WD®	25% dinocap	1-10 mg/L
Orthocide®	50% captan	1-10 mg/L
Kelthane 35®	35% dicofol	1-10 mg/L
Bayleton®	5% triadimefon	50-200 mg/L
Cuprox®	50% copper oxychloride	50-200 mg/L
Roval®	20% triforine	50-200 mg/L
Karmez®	80% diuron	50-200 mg/L

From Conner (1983)

hexyl isocyanate > butyl isocyanate = methyl thiophanate > benomyl > iprodione (degrades to carbendazim and butyl isocyanate). They found no inhibition from 3-butyl-2,4-dioxo-S-tirazino[1,2-a] benzimidazol (STB). They also noted that 3 to 30 μM Captan® strongly inhibits the O_2 uptake of the yeast.

Another fungicide used in Europe, Euparen®, is also a potent yeast inhibitor. Ronilan® and other systemic fungicides have not shown any significant inhibitive properties toward yeast. The systemic fungicides are probably less likely to be a problem than contact ones. The breakdown products of these materials can possibly be a concern for the future. One example of this is Orthene-50®. Under the reducing condition of yeast fermentation and wine aging, it breaks down to CH_3-SH and CH_3-S-S-CH_3, both unpleasant compounds. It is not allowed for use on grapes, even though it is an excellent pesticide from other standpoints.

Yeast inhibition can come from fatty acid produced by the yeast, according to Larue and Lafon-Lafourcade (1989). In particular, octanoic, decanoic and dodecanoic acid can slow fermentation and lower the viable yeast population. The esters of these acids can also contribute to the adverse effects. Addition of about 0.29 g/L of ghost cells (yeast cells that have been treated to make yeast autolysate), after the fermentation has proceeded about 25% to completion, will then adsorb the offending acids and esters. This protects the yeasts from the acids' inhibitory effects. The addition of 300 mg/L of ghost cells or microcrystalline cellulose was found by Minárik and Jungova (1988) to adsorb some inhibitory pesticides. It also aids the fermentation of high sugar wines.

Arsenic compounds have been used in certain areas as a pesticide. While it is not illegal if used correctly, the wines made from treated vineyards can have residual amounts of arsenic. Other treatments should be substituted for the use of arsenic sprays.

In a review concerning effects on the fate of pesticides in grape and wines by Cabras et al. (1987), a few general comments were made concerning the use of vineyard fungicides. Most fungicides do not penetrate the soil deeply and are little worry for contamination of water supplies. Persistence in the soil varies with compounds and with soil flora and organic materials. Residues in grapes from sprays depends on many factors—spray concentration, weather, type of compound, etc. Clarification of the must, bentonite and carbon treatments will lower the levels of most of the fungicides. Seldom is the sensory quality of the wine affected by the fungicides. Only a few will inhibit yeast growth.

While the ultimate legal responsibility for problems arising from grapes treated in vineyard rests with the farmer, the winery must share the responsibility. If the wine is contaminated, then the winery's name bears the bad publicity and the lawsuits that may follow. Therefore, care in this matter is important. The analyses of the juices (or wines more properly) should be attended to. If the regulations concerning the amounts of times between application and harvest are properly adhered to, then there is little danger of a problem. However, the winery may not always have control of this.

YEAST GROWTH IN AND FERMENTATION OF JUICE

Growth

The main factor that concerns yeast growth in grape juice is yeast nutrients. Of these, the amino acids are very important. The sugar level, even at °Brix of 25 to 30, has little effect on growth. The pH in the ranges, usually found in juices 2.75 to 4.20, has little effect except as a factor to increase the inhibition of SO_2 additions, if at critical levels. In rate of fermentation studies on several juices, the nitrogen content could be correlated well with overall fermentations and cell growth (Ough and Kunkee, 1968). Studies of numerous authors have shown the dependence of growth on amino acid content. The paper of Ingledew and Kunkee (1985) demonstrated this under laboratory conditions. The effect of assimilable amino acids, added to juice previously having sluggish fermentation, finished quite rapidly at reasonable fermentation temperatures. This effect of lack of nutrients on fermentation has also been well demonstrated by Bell et al. (1979) and Ough and Nagaoki (1984) in field experiments. An example of the dependence on a α-amino acid content for the velocity of the fermentation is seen in Figure 3-2 for Cabernet Sauvignon grapes fermented under similar conditions.

The need of aeration, also pointed out by Ingledew and Kunkee (1985), is important. In juices with insufficient α-amino acids, In-

FIGURE 3-2. Changes in fermentation rates in red wine related to the initial α-amino acid concentration in the must.

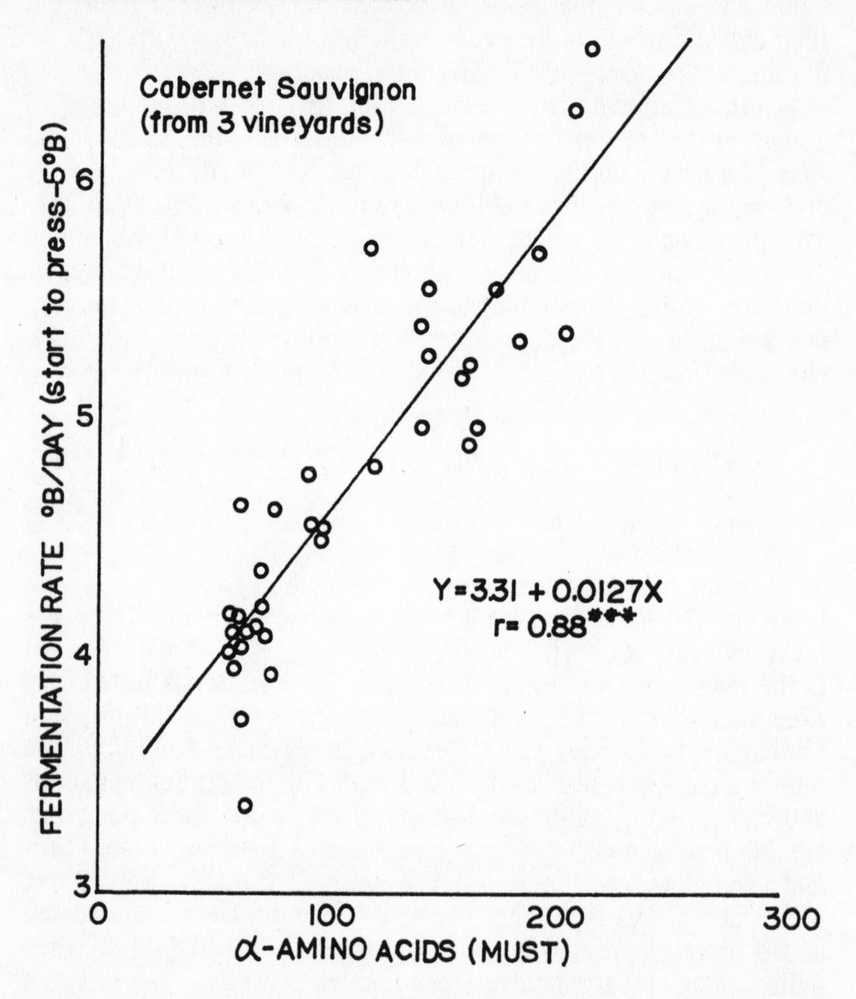

gledew et al. (1987) have suggested oxidation of the juice during the yeast growth period so the yeast can use proline as a substrate. Then cells in sufficient number to finish a possible stuck fermentation would result. There is an absolute oxygen need for the two enzymes required for the metabolism of proline to glutamate. The disadvantages may outweigh advantages because of possible damage to the quality of the wine. Blending with a juice high in amino acids is a better solution, or addition of diammonium phosphate (DAP).

In small laboratory trials, adequate oxygen is present unless care is taken to avoid it. Commercial fermentations have less chance of air contact because of much smaller surface to volume ratios. It is probably wise to aerate the juice for yeast viability considerations. One should realize this aeration should occur *after* inoculation, as the polyphenol oxidase (PPO) will rapidly scavenge the oxygen. In fact, if the enzymes are not inhibited by SO_2, it would be a competition between the yeast and PPO for use. The rapidity which the PPO used oxygen would probably favor that pathway. There may be other factors which are counter-indicative of this treatment, however. These include urea accumulation, as well as flavor losses.

The need for the sterol concentration to be sufficient to maintain the resting cells is essential (Larue and Lafon-Lafourcade, 1989). The problem arises when the amino acids are low also. It can occur when a sufficient and good crop of yeast grows initially. The sterols in grape are ergosterol, cholesterol, lanosterol and oleanolic acid (grape wax). The need for these is especially pronounced with high sugar grapes, well clarified and fermented cold. Under these conditions, the time for resting cell activity can be very long.

Jones and Ough (1985) measured the weight of cells produced from Cabernet Sauvignon juice from vines grown under different conditions – one with high nitrogen content and the other low. The weight of the dried cells grown on the rich medium were compared to those grown on the low nitrogen juice. The differences in the ethanol produced could be accounted for by the differences in nitrogen in the juices. The high nitrogen grapes produced more cells and used a larger percent of the carbohydrate for that purpose.

Vitamins are not usually scarce in grape juice. The dependence of some common yeasts on these vitamins was investigated (Ough et al., 1989). Only if biotin was carefully removed could any effect be shown on yeast growth. The only other vitamins which had any growth requirement by the yeasts used was pantothenic acid. The levels showing inhibition again were well below any levels found normally in grapes.

The temperature speeds the growth of yeast and the velocity of the fermentation. A maximum temperature of about 40°C is reached, then the cells begin to die. A few degrees before this metabolism begins to change, these temperatures are well above normal wine temperatures. White juices are inoculated and ferment from 10°C to 15°C, normally, and red musts from 20°C to 30°C. Arrhenius plots of fermentations over this wide range, 10°C to 32°C, show good linearity.

There is no question that ethanol inhibits fermentation and yeast growth. Ough (1966b) demonstrated that increasing temperatures causes more rapid cell death at increased alcohol levels. This was further discussed by van Uden (1989a) with similar conclusions. The effect of ethanol on rate of cell growth was discussed by Pamment (1989). He concluded that alcohol inhibition of yeasts is noncompetitive and a modification of the Monod expression of cell growth is appropriate.

$$\mu = \frac{\mu^{\circ}\max S\, f(p)}{K_s + S}$$

μ = h^{-1} (rate of growth/unit mass × amount of biomass)

$\mu^{\circ}\max$ = (maximum sp. growth rate without inhibitors)

S = (concentration of growth-limiting medium)

K_s = (saturation constant)

The equation can be rephrased

$$\frac{1}{\mu} = \frac{K_S}{\mu_m^p S} + \frac{1}{\mu_m^p}$$

μ_m^p = (maximum specific growth rate in presence of p, ethanol)

This equation is valid for yeasts when K_s is not changed by inhibitor concentration.

Jones and Pierce (1964) determined the order and rates of amino acid uptake in *Saccharomyces cerevisiae*. They divided them into four groups. The most easily assimilable group consisted of arginine, asparagine, aspartic acid, glutamic acid, glutamine, lysine, serine and threonine. Only two other of the amino acids were found in relatively large amounts in grape juice – alanine and proline. Alanine is in the third group and proline in the fourth by itself. Egbosimba et al. (1988) found essentially the same order with *S. cerevisiae* from malt extract medium.

In grape juice, arginine, when in excess, is metabolized slowly after the rapid uptake of ammonia, serine, threonine, glutamic acid, glutamine, aspartic acid and asparagine. It is used more slowly than alanine.

Obviously, the makeup of the amino acids available will have a bearing on growth of the yeast. Some of the major differences in the amino acid content of grape juice can be accounted for by cultivar differences. For example, Cabernet Sauvignon is relatively low in arginine and high in proline while Pinot noir is higher in arginine and lower in proline. Likewise, other differences in the cultivars, such as Sauvignon blanc and Chardonnay are evident. These differences can be overshadowed by viticultural practices that can cause excess or very low levels to occur.

Amino acid content of free-run juice from the crusher, from juice from the drainer tank, after a few hours contact also being pumped over and the juice from the press can vary greatly. The free-run juice can be relatively low, while the juice after skin contact can double in α-amino acid content. The press juice can have even higher levels. For this reason, many experienced winemakers say if one is having trouble with a slow fermentation, add some press juice to it.

If wines are low in the easily assimilable α-amino acids, then diammonium phosphate is added to supplement. Ammonia is easily assimilated by the yeast; the usual amount is about 1/2 pound/1000 gal. Rozes et al. (1988) indicated aeration and 25 mg/L of added ammonia could give increased number of viable cells and faster fermentations. Twenty to thirty cc/L of oxygen gave maximum results with biomass increases of as much as threefold. Ninety percent viability was maintained with these oxygen levels (Sablayrolles and Barre, 1987). If there is any doubt about the fermentation being slow, or possibly sticking, both treatments are advisable. Experience from certain vineyards from previous years can indicate a problem.

Fermentation

van Uden (1989b) discussed the effect of ethanol on the glucose transport system. He concluded that the alcohol noncompetitively bound to the membrane proteins, thus effecting the transport. He suggested the ethanol inhibited the membrane proton pump.

Pamment (1989) found that *all* models to predict fermentation rate were poor. They could only be reproduced in the substrate the model was derived from. The problem is defining the cell growth and other inhibitor effects that are specific for the particular substrate.

The final ethanol content in the wine depends on many nutrient conditions and the temperature of fermentation. The vapor pressure of the alcohol in solution increases with fermentation temperature; thus a larger percentage is lost in the fermentation as temperatures rise. Estimates of these losses have been made by Williams and Boulton (1983). Figure 3-3 indicates expected losses.

The level of insoluble solids in the white juice is critical to the yeast fermentation at lower fermentation temperatures. Normal settled and racked or centrifuged juices have about 1-2% solids. This is sufficient to furnish attachment points for yeast. The yeast then will ride these solids and the CO_2 given off will attach to the solid enough to cause it to rise. When the bubble is large enough, it will release and yeast and particles will float down and repeat the performance. Yeast grown in juices with less than these levels of solids tend to pack on the bottom. This makes the surface available for sugar transport minimal. Diatomaceous earth or bentonite can be substituted for the grape solids with similar results. Photomicrographs (Ought and Groat, 1978) of the material show the cells adhering (Figure 3-4). Stirred samples without solids ferment as well as the ones with solids added.

The yeast cultures should be observed microscopically to be sure that the yeasts are of the required type and not oxidative yeasts and bacteria. Valuable experience can be acquired by learning to identify the yeast in the pure culture. When fermentation problems arise later, differences between good wine yeast and spoilage yeast or bacteria will be distinguishable. Depending on the temperature, signs of fermentation should occur between 24 and 48 hours after

FIGURE 3-3. The linear relationship between the reciprocal absolute temperature and the logarithm of the total alcohol lost for the initial sugar indicated (From Williams and Boulton, 1983).

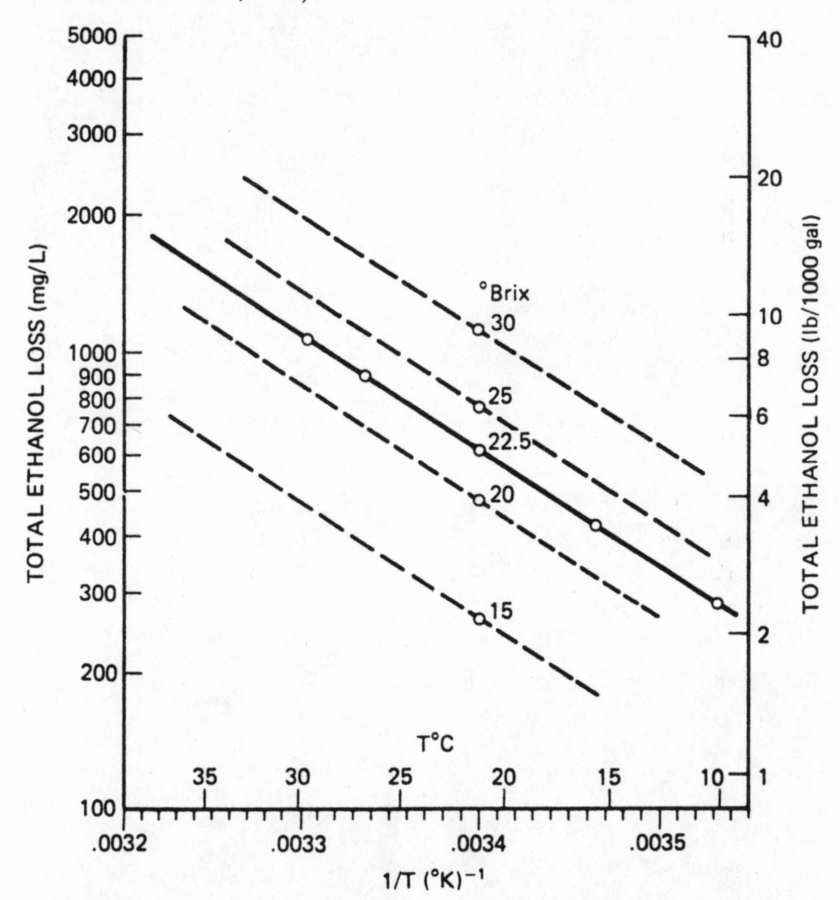

the must is inoculated with the yeast culture. The white juice should be noticeably more turbid with gas bubbles starting to form (Figure 3-5). With red grapes, the rise of the cap (mixture of crushed grapes and gas bubbles) is usually evidence that the yeasts are growing and fermentation has begun. Figure 3-6 shows the time-temperature relationship for starting times. The variety of the yeast chosen should be from the *Saccharomyces cerevisiae* species. Several different races and strains are available. Included in these are Montrachet,

FIGURE 3-4. A. Yeast from an unstirred fermentation, B. Yeast from a stirred fermentation, C. Yeast unstirred but with 1.5% grape solids added, and D. Yeast unstirred with diatomaceous earth added.

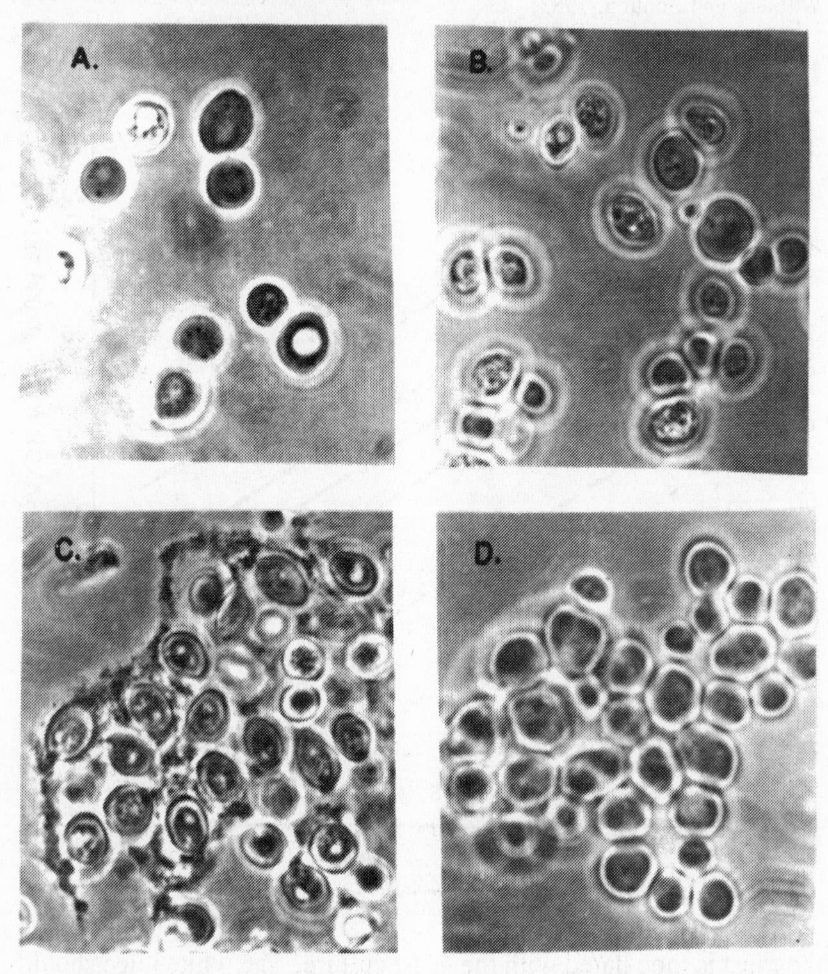

Epernay II, California Champagne, Prise de Mousse and many more in dry yeast form (in USA).

After crushing, draining or centrifuging, settling and racking any needed adjustments of acidity (or sugar if permitted) are taken care of. The yeast culture is added. The amount should be sufficient to

FIGURE 3-5. A small scale fermentation showing initial foam and cloudiness. Note the fermentation lock which excludes air.

FIGURE 3-6. The linear relationship of the reciprocal of the starting time (hours) to the temperature of the medium, for two separate years. 1963 data are averages of 10 fermentations at each temperature; 1964 are averages for 12.

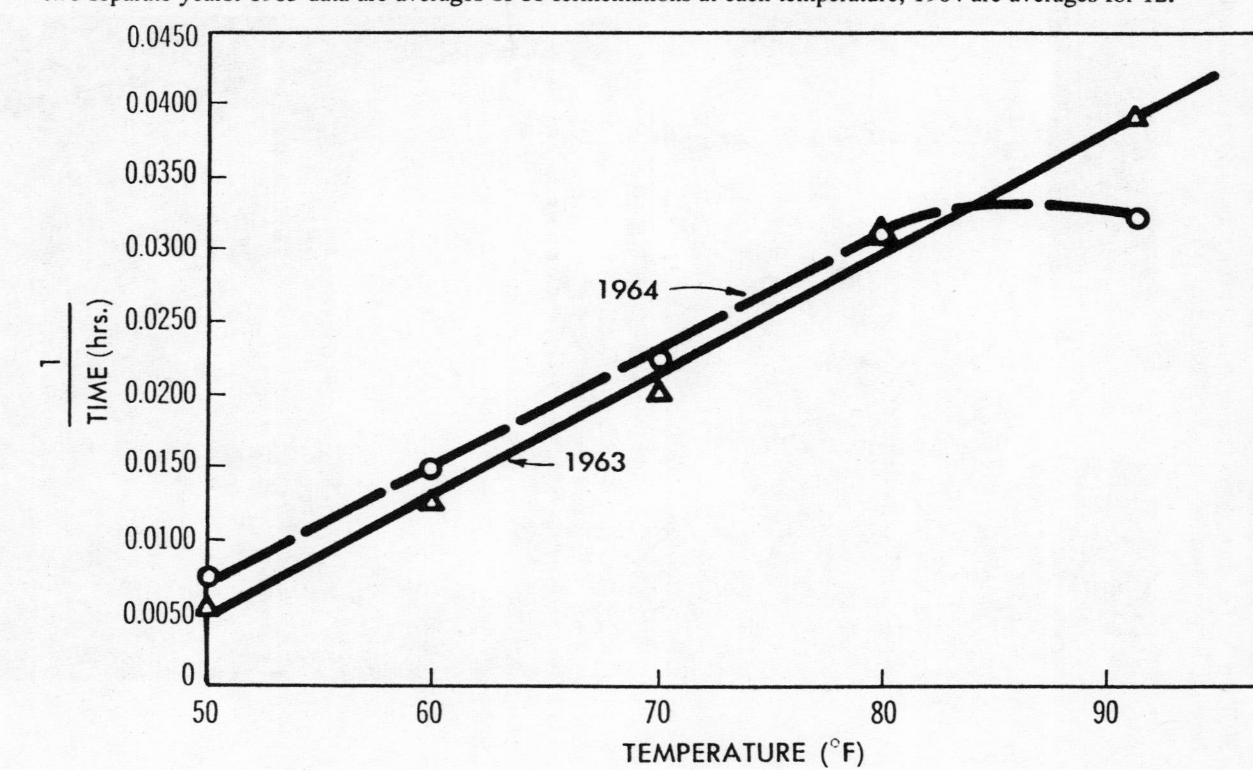

give 10^6 cells/ml or more. The rate of fermentation will depend on many variables including the number of viable cells that grow. This, in turn, depends on the nutritional constituents of the juice, sugar and alcohol concentration, and pH, temperatures and the degree the yeasts are agitated.

Temperature

When the temperature is elevated above 20°C, the yeast multiply very rapidly and also a more rapid fermentation occurs (Ough, 1966a). With temperatures from 25-30°C, the fermentation takes only two to four days, usually. Under these conditions, white wines lose varietal character and develop adverse flavor which correlates with increased temperature. During fermentation, yeasts produce esters which have a fruit-like character (short chain esters) and others which have a heady aromatic quality (longer chain esters). The very fruity esters accumulate in the wines fermented cold.

Figure 3-7 shows typical concentrations of amyl acetate over the fermentation period of a juice fermented at several different temperatures. Other individual esters respond similarly. The longer chain, higher boiling esters accumulate more at the higher fermentation temperatures. The lower weight acetate esters contribute strongly to the so called "fermentation bouquet" in freshly fermented white juices. This is especially true of juices relatively free of solids and fermented at cold (10-12°C) temperatures. Ramey and Ough (1980) followed the hydrolysis kinetics of these esters. Figure 3-8 shows the relative temperature effect on the pseudo first order kinetics of hydrolysis. The formation of the esters is incidental to the yeast growth, but may have an indirect effect of yeast survival. Drosophila are strongly attracted to the odor of the esters and could transport the yeast on their legs and bodies from place to place.

In synthesis of lipids, proteins and other organic compounds, the formation of acetyl CoA is essential. This compound reacts with alcohol to give esters.

$$R_1\text{-}\overset{\overset{\displaystyle O}{\|}}{C}\text{-SCoA} + R_2OH \longrightarrow R_1\text{-}\overset{\overset{\displaystyle O}{\|}}{C}\text{-}OR_2 + \text{CoASH}$$

FIGURE 3-7. Effect of temperature and progress of fermentation on concentration of isoamyl acetate. 10° C = ⬡ , 15° C = ○, 20° = △ , and 30° C = □.

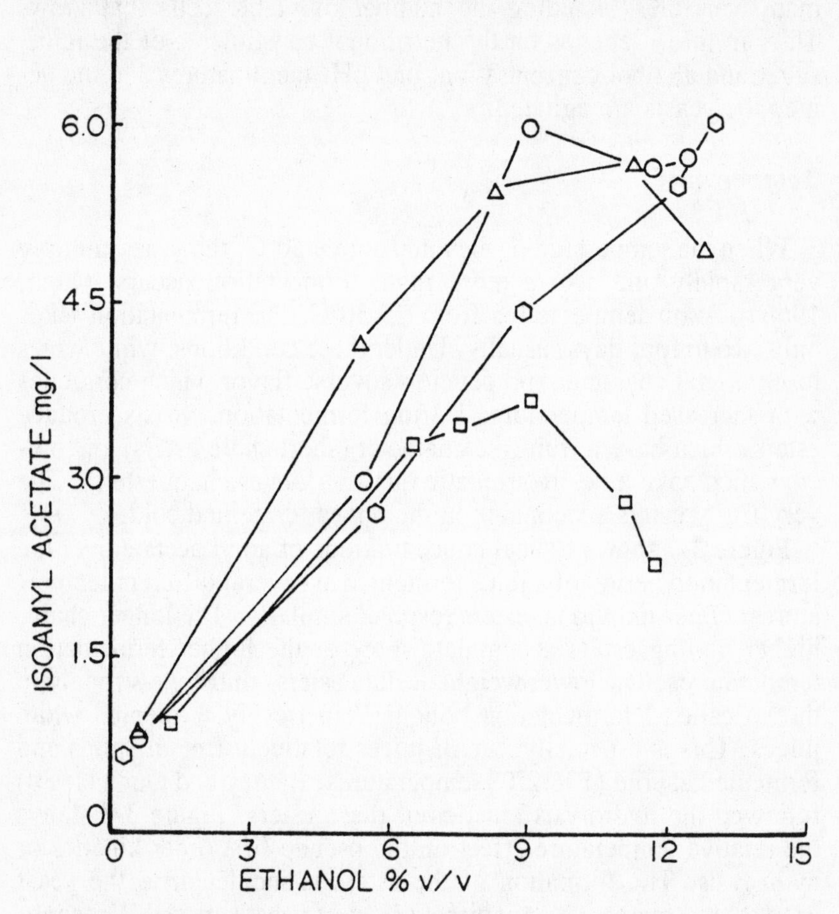

The formation of these esters is driven by the supply of energy-rich acetyl-SCoA. The esters formed will exceed the equilibrium amounts. The rate of hydrolysis is relatively slow at lower temperatures. The *Saccharomyces* sp. have relatively minor amounts of esterases to speed the hydrolysis. Pectic enzymes contain esterase, though no reports have been made on how these affect these esters.

Contrary to the acetate esters, the ethyl esters of tartaric acid form during aging and not by activity of the yeast. They contribute

FIGURE 3-8. Effect of temperature on the hydrolysis of isoamyl acetate.

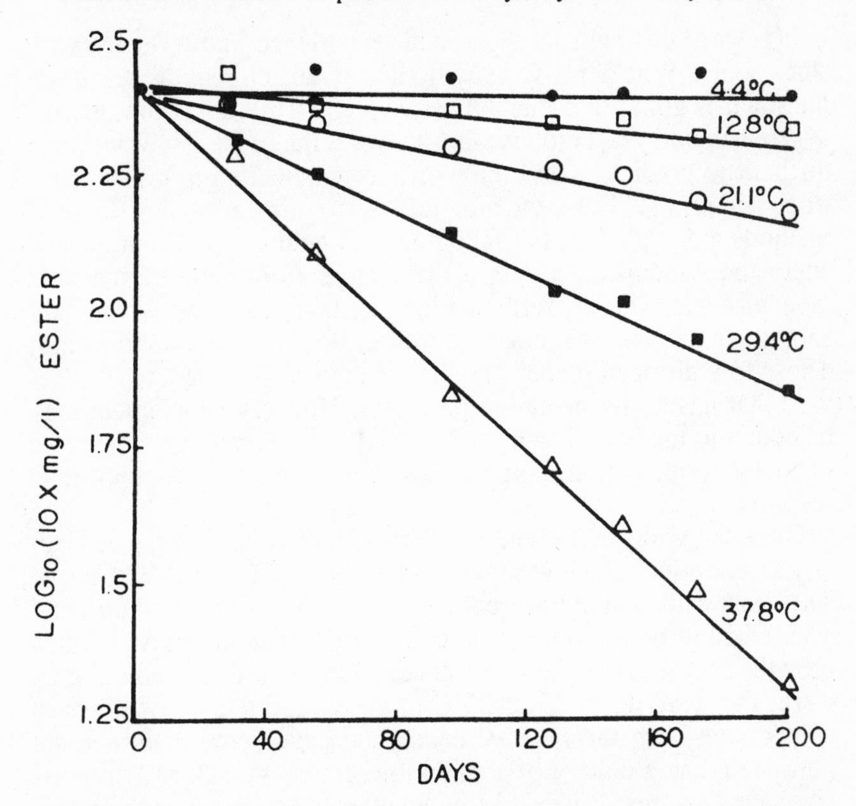

nothing to the odor of the wine (Edwards et al., 1985). Also, they noted that the reduction in acid by the esterification tended to mellow the wine. They detected no diethyl tartrate in their work. Formation of the ester followed second order kinetics, while the hydrolysis was first order. The net formation rate was slow.

Varietal character, such as the terpenols in White Reisling and fruitiness as in Chardonnay, are better preserved at lower fermentation temperatures. Commercially, many of the quality white wines are fermented at 12-15°C. At this temperature, the good esters and varietal character are preserved and the yeast can grow properly, fermenting the wine in a reasonable length of time. All fermentations should be completed in 20-30 days.

Stuck Fermentation

It is very difficult, as those with experience know, to restart a stuck wine. Whether a wine will stick (stop fermenting before all the sugar is gone) or not is not easily predictable. Lack of nutrients to grow a good yeast crop is almost always the cause. However, it is difficult to predict, unless the winemaker is willing to do some nitrogen analysis. The best measure is α-amino nitrogen. Several methods are available, but take time and equipment. From experience, the winemaker may see that grapes from certain vineyards give problems during fermentation. In that case, the winemaker should encourage the vineyard owner to fertilize the vineyard. Those low nitrogen grapes may be blended with those from a vineyard that has no fermentation problems. Nitrogen supplement may be added to the juice [such as $(NH_4)_2HPO_4$]. White wines, because of lower fermentation temperatures, are more likely to stick than red wines.

Once the wine sticks, the problem of restarting is serious. Usually a wine sticks at a low sugar level – say 10 g/L and 12% ethanol. The yeast cells that are present have used up most of the nutrients. The ethanol concentration is in a range that strongly inhibits growth. Not a lot of options remain. One possible method is to warm the wine up to 25 or 30°C to encourage release of nitrogen compounds from the inactive cells. Then add some diammonium phosphate and a culture of alcohol tolerant yeast such as *S. cerevisiae* race *bayanus* (Prise de Mousse) after lowering the temperature to 20°C. Lastly, give the wine a quick aeration just after adding the new culture. Sometimes, after a month or two, wines will restart and finish without any treatment. Wines that do stick are of inferior quality to those that do not stick.

Hydrogen Sulfide

Hydrogen sulfide (rotten egg smell) can be a problem if free sulfur is present in the juice or the nitrogen level is low. Settling and racking, or centrifuging, will take care of the free sulfur problem in white juice. In red grapes, aeration during and after fermentation helps. The use of dusting sulfur in vineyards should cease six weeks before vintage. Problems of H_2S, due to low nitrogen, can be alleviated by additions of diammonium phosphate at 500 mg/l or 4.17 pounds/1000 gals. The reason for formation of H_2S from the free sulfur is the reducing conditions during fermentation. This can easily change elemental sulfur to sulfide. The pH is such that most of the $S^=$ is in the H_2S form.

$$H_2S \rightleftharpoons H^+ + HS^- \rightleftharpoons S^= + 2H^+$$

The case of lack of nutrients is covered further on.

Following the Fermentation

Measuring the change in the °Brix of the fermenting musts is the most satisfactory method of following fermentation progress. Usually, if the data are plotted against time, an S-shaped curve results. Figure 3-9 shows a typical series of curves for juices fermented at different temperatures. The end final °Brix, when the wines reach

FIGURE 3-9. Changes in the time to ferment a juice as affected by fermentation temperature.

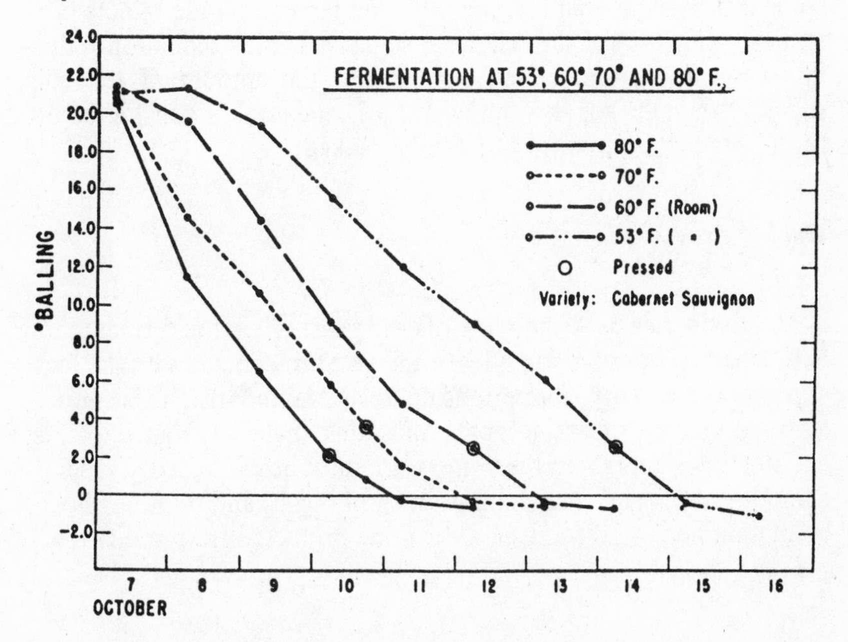

dryness, are always less than 0°Brix because of the effect of ethanol.

The more common practice is to record the °Brix. When it is constant and negative for several days, the wine can be racked. The °Brix values for dry wines depend on the original amount of sugar because the alcohol produced influences the °Brix reading. Alcohol has a density less than water. When the wine metabolizes the sugar and forms alcohol, the °Brix changes to a greater extent than if the effect was sugar loss alone. If the °Brix in the juice was around 20°Brix, then the °Brix of the wine when dry will be perhaps −1.2°Brix. If the °Brix in the juice was 24°Brix, then you might expect a value of −2.3°Brix at dryness. There is no formula or set rule for determining dryness. Each wine can have differing amounts of residual compounds which will affect the specific gravity. A refractometer cannot be used for this purpose for obvious reasons.

To determine if a wine is dry, check the fermentation by use of the Ames® sugar tablet or analyze for reducing sugar. This test will indicate how much residual sugar remains in the wine. (The Ames® pill test is good only in amounts up to 20 g/L or 2%. Directions and color comparison charts come with the tablets. These tablets contain the reagents for the reducing sugar test, plus oxidation-reduction color indicators which correlate to the amount of reducing sugar.) When the sugar is below 1-2 g/L, the wine is considered dry for all practical purposes and can be racked.

Racking the Wine

Most wineries rack the wine off the yeast when the fermentation is finished. However, some leave the yeast in contact with the lees, stirring occasionally, for periods up to several months. This encourages the release of amino acids and other material from the yeast and imparts flavors into the wine. Care must be exercised that spoilage does not occur. The main protection is SO_2 and the free should be maintained at 20 mg/L and the wine protected from oxidation.

White Wines

During the fermentation of white wines, there is little to do but monitor the progress and temperature. In some special cases, other treatment of white juices may be considered other than cool fermentation of the expressed juice.

Marais (1987) looked at the wines made from different heating and skin contact times. He suggested that skin contact at 10°C, followed by a certain degree of heat treatment of the juice, gave typical Gewüztraminer wines with fruity, estery, spicy characteristics. It should be noted that different climates, crop levels and other factors can modify the results.

Some winemakers ferment Chardonnay juices at 20°C, or slightly above, in 50- or 55-gallon oak barrels. This procedure gives wines with increased wood extracts and flavors. These wines easily undergo malolactic fermentation as they usually have had some skin contact. These may also have extended lees contact to further enrich flavors.

Botrytized Wine

Wines made from botrytised grapes include some of the fine German sweet table wines, Tokays from Hungary and Sauternes from France. Equally fine wines have been made in California, South Africa and other places that occasionally have the right climatic conditions. These conditions demand a few warm humid days for the *Botrytis* to grow and penetrate the fruit. Then, a few cool dry days are needed for the mold to dehydrate the berries. Sometimes this is helped by some drying after picking. Fructose is a more desirable sugar to remain in these wines than is glucose because it is about one-half again as sweet. Lemperle et al. (1988) looked again at the glucose/fructose ratios of grapes and wines. The fruit, when ripe, had ratios from 0.90 to 1.06. The ratios either drop or remain fairly constant over the maturity range. Using one yeast, they found delayed use of fructose towards the end of the fermentation. Other yeasts used up the fructose more completely at the finish of the fermentation. In looking at dry or near dry wines and sweet table wines, the values and ratio are shown in Table 3-3. The sparing of

Table 3-3. Reducing Sugars and Glucose and Fructose Ratios in German Wines.

Type[1]	Reducing sugar g/L	Glucose g/L	Fructose g/L	Glucose/fructose g/L
Table				
Quality	3.76	1.02	1.33	0.77
Kabinett	3.32	0.76	1.09	0.70
Spatlese	4.64	1.20	2.00	0.60
Sweet Table				
Quality	14.55	6.34	7.85	0.81
Kabinett	18.28	7.57	9.76	0.78
Spatlese	36.25	16.22	18.38	0.88
Auslese	57.79	18.31	34.26	0.53
Beerenauslese	141.65	40.73	88.40	0.46

[1]The mean of 10 wines in each category.

fructose is most noticeable in the sweeter wines. Other reducing sugars are noticeably present in these wines, besides glucose and fructose.

Nelson and Nightingale (1959) studied the commercial production of natural sweet wines by artificial application of *Botrytis* spores onto the grapes under controlled conditions. The product was excellent, but the cost was high.

Red Wines

To extract the maximum color and flavor from the skins of red grapes, they must be mixed in some manner. Different ways have been suggested – pumping over, submerged cap, stirring and roto tanks – for this extraction. The method of choice for most wineries is pumping over twice daily with about half the tank volume pumped over the cap each time. The cap must be kept thoroughly wet to prevent bacteria and other alcohol non-tolerant molds and yeasts from growing on the surface. Most tanks used are now closed tanks and the lid opened only to pump-over. Tank sizes range from 500-50,000 gallons for most table wine wineries. The temperature of fermentation ranges from 20°C to about 32°C. The temperature of fermentation has considerable effect on the color and phenols that are extracted. Increasing temperature causes increased amounts of both in the resultant wine. Likewise, the ethanol content influences the amount of color extracted and retained in the wine. Sulfur dioxide also influences the amount of color in the wine. Lastly, the degree of severity of the pressing also influences wine color.

The temperature in the skins in the cap of the fermenting wines can go 20°C or more above the fermentation temperature of the liquid. This is due primarily to the yeast that resides in the skins just above the liquid level. They give off heat that is trapped in the cap. The pumping over does mediate these temperatures, but only for a short time. Fermentation temperatures in the liquid above 32°C can cause excessively high temperatures in the cap, induce the thermophylic bacteria activity and lead to wines with very disagreeable odors. Submerged cap wines have less color than those that are pumped over. Temperatures in the cap can also go fairly high. The CO_2 and yeast become trapped in the cap area. The liquid is either not in contact with all the skins or is not circulating to remove the heat. Figure 3-10 is a seldom-seen example of an open-tank, submerged-cap, red wine fermentation. Stirring is destructive to the skins causing later clarification problems. Roto tanks are more expensive (Figure 3-11). They are medium-sized revolving tanks with internal vanes. When the tanks are rotated on the horizontal axis, the juice and skins are mixed well. Good color extraction results. Another type used in South Africa is a pressure system. The fermenting wine in the tank is allowed to build up pressure, then an automatic valve releases the built up CO_2. This causes the must to be vigorously mixed by the escaping CO_2 from the liquid. Color extraction is not as complete as pumping over.

Pinot noir sometimes is fermented with its stems present or even ground up. The stems, if reasonably dry and from mature fruit, impart a peppery aromatic taste to the resulting wine. In previous

FIGURE 3-10. An example of an open tank, submerged cap fermentation. This is rarely seen now (Henschke Winery, Barossa Valley, Australia).

FIGURE 3-11. Banks of rototanks at R. Mondavi Winery, Oakville, California.

times, it has been known for winemakers to add a few leaves of eucalyptus to enhance the "pinot" character. This aroma, while pleasant if not overdone, is not the aroma of Pinot noir.

Red wines seldom have any problems going to dryness. Reasons for this are the extra source of nutrients from the skin and pulp and the more elevated fermentation temperatures.

When the fermenting wine is to be separated from the skins is a choice for the winemaker. Long contact time, high ethanol content and more grape cell breakdown allow release of phenols, color and flavor (see Figure 3-12). Also, the longer the skin contact, the longer the wines will take to mature. Wines pressed at 10-15°Brix mature in a very short time, while those kept on the skins until near dryness require much longer barrel or cask aging. Quality wines are now pressed about 1-2°Brix (Cooke and Berg, 1983). In addition, the degree of press determines, to some extent, the amount of color, phenol and flavor in the wine. The hard press juice is kept separate for later blending. The amount varies from 10 to 80 gals/ton out of a possible wine yield of 160-195 gals/ton of grapes. When the wine is pressed, it is usually finished fermenting as a white wine.

Maceration Carbonique

When whole grape clusters are placed into an anaerobic atmosphere of CO_2, there is a strong effect on the internal metabolism of the berries. The malic acid is converted to ethanol. According to Flanzy et al. (1987), glucose combines with CO_2 to give malate, aspartate will be converted to fumaric and then to malic and succinate. Glutamate was indicated to form α-aminobutyric acid. In their ^{14}C investigation, the shikimic acid pathway gave several volatile phenolic compounds. These were associated with the special aroma found in these wines.

For effective operation of this type of fermentation, the clusters are placed whole in the tank. The air is excluded either by CO_2 gassing or the addition of a small portion of a fermenting must. The first phase of the fermentation takes place at a tank temperature of 30-32°C. The weight of the berries and the metabolic activity and the ethanol breaks down the cells of the berries and color is extracted. After about 8-11 days, the berries are ready to press. The

FIGURE 3-12. Relationship of the log of the tannin, color, alcohol to log of time of fermentation.

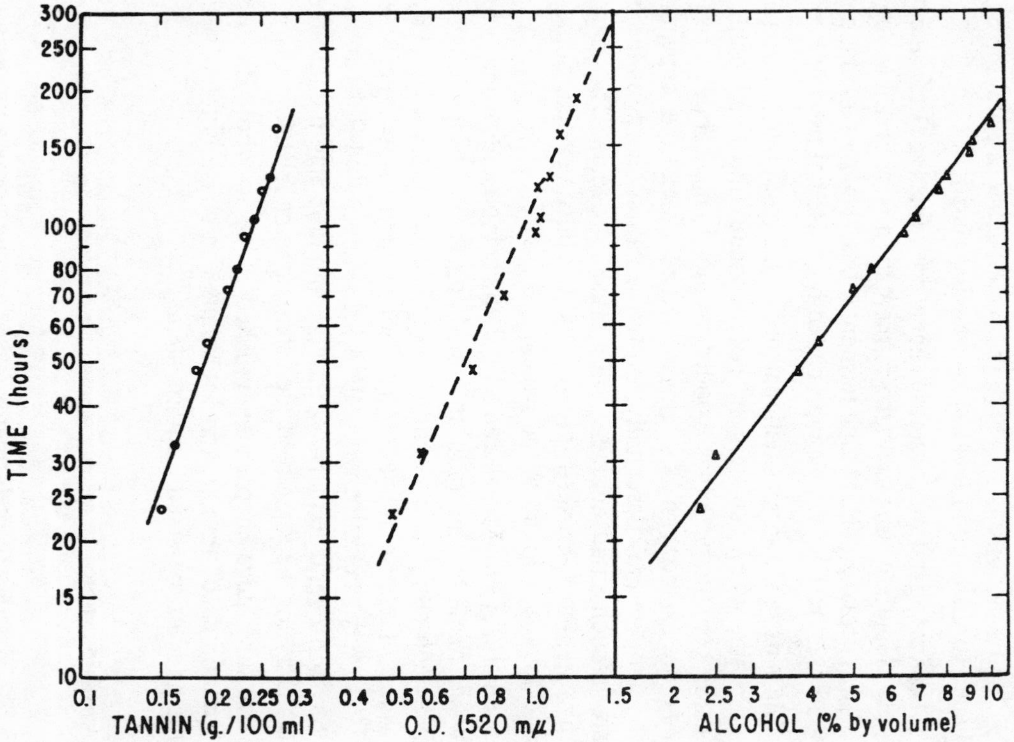

press juice, along with the free-run, are combined and fermented on to dryness at 18-20°C. The malolactic fermentation is finished then or before the yeast fermentation is completed. As soon as the microbiological activity is finished, the wine is checked and the SO_2 adjusted to the desired level, and the wine clarified.

Less refrigeration capacity is needed for this type of fermentation as the length of time to do the fermentation is extended beyond that of the normal crushed grape fermentations. The flavors that result in the wine are not always appreciated by those not used to them.

In Australia, some smaller wineries harvest their red grapes and store the whole clusters in polyethylene film-lined containers. These containers, either wooden or metal, hold about a ton. The bags are gassed with CO_2, closed, and held for a time, depending on the temperature. Essentially, no fermentation occurs as the grapes are in such small amounts, little crushing occurs. The grapes are then stemmed, crushed, pressed and the juice yeasted. Some berry character was found in the wines.

Ducruet (1984) measured 84 volatile components from "Merlot" wines made by the traditional method, "maceration carbonique." Nineteen components exhibited significant differences. Only four were in higher amounts by the "maceration carbonique" method. These were benzaldehyde, ethyl salicylate, ethyl-9-decanoate and vinylbenzene. The pathway through shikimic acid is indicated as a pathway for the three benzene-related compounds.

Adding juice or partially crushed grape to whole clusters and fermenting to simulate this technique can cause bitter wines from the phenols extracted from the stems.

Rosé Wines

Rosé wines were fermented as white wines as soon as they are pressed or drained. This preserves the fruity ester aromas and prevents coarse flavors due to elevated fermentation temperatures. Likewise, "blush" wines are treated as white wines.

FERMENTATION BIOCHEMISTRY

Glycolysis

The pathway from glucose to alcohol is an 11-step enzymatic degradation with the end products mainly ethanol and CO_2. Fructose requires one less enzymatic reaction. Intermediate in the pathway, fructose-1,6-diphosphate is split into two three carbon compounds (D-glyceraldehyde-3-phosphate and dihydroxyacetone phosphate). Another enzyme converts part of the dihydroxyacetone phosphate to D-glyceraldehyde-3-phosphate. The remainder of the dihydroxyacetone phosphate is converted irreversibly to glycerol via a two-step reaction. The D-glyceraldehyde-3-phosphate is then converted through seven more enzymatic reactions to ethanol. What purpose does this conversion achieve for the yeasts? Little that happens in nature is without reason. Consider the adenosinetriphosphate (ATP), the energy-rich compound which drives many reactions. There is a loss of two ATP for each molecule of sugar phosphorylated to fructose-6-diphosphate. Essentially, four ATP are returned to the yeasts when the 1,3-diphosphoryl-D-glycerate is converted to 3-phosphoryl-D-glycerate and the phosphorylenolpyruvate converted to pyruvate, an approximate net gain of two ATP. The production at the beginning of fermentation of NAD^+ by the glycerol side reaction is essential. It allows the phosphorylation of D-glyceraldehyde-3-PO_4 to start producing NADH. This is needed for the final step in converting acetaldehyde to ethanol. Measurements of acetaldehyde in juice early in the yeast growth phase of the fermentation show rather high levels of acetaldehyde, but the concentrations drop off dramatically as the fermentation starts in earnest. The only step where CO_2 is given off is when pyruvate is converted to acetaldehyde. This step requires the cofactor thiamine pyrophosphate. Other minerals required at various steps are Mg^{++}, Zn^{++}, Co^{++}, Fe^{++} and Ca^{++} or K^+.

The amount of glycerol formed varies with conditions. For example, the amount increases with fermentation temperature (Ough et al., 1972).

The amount of sugar converted to ethanol depends on the needs of the yeasts and the balance of nutrients. If many amino acids need to be synthesized, greater amounts of fusel oils will be formed. If amino acids are low, then less yeasts will be made and more sugar converted to ethanol. The very complicated biochemistry of the yeasts make any absolute predictions of total cell growth or ethanol production difficult.

Fermentation is an exothermic reaction giving off around 24Kcal per mole of glucose. The rate of fermentation (dependent on yeast strain, temperature of fermentation, number of cells, sugar content, alcohol content) regulates amount and rate of heat formed. The size of the tank, surface to volume ratio loss through tank walls, and other factors all have a bearing on the amount of heat being retained in the form of higher temperature of the wine. Having sufficient cooling capacity to handle the worst case situation is essential for any winery. If tanks "get away" and become very warm, control is difficult, if not impossible, and wine quality suffers.

Other Pathways

Many of the biochemical pathways for the major components within the yeasts are known. References to major texts on biochemistry are given in the References and Bibliography. Some other specific areas of interest are the synthesis of SO_2, higher alcohol fermentation, some key amino acid synthesis and metabolism formation of acetoin, diacetyl and 2,3-butanediols, acetic acid and succinic acid. Farkaš (1988) suggests that about 10-12% of the sugar present is diverted to yeast biomass and side products. His average values were 2.5% for yeast biomass, 4.2% glycerol, 0.3% succinic acid, 0.1% lactic acid, 0.1% for 2,3-butanediol and 0.1% acetic acid.

The formation of SO_2 from sulfate by yeast is a four-step reaction. It requires two ATP per molecule of SO_2 formed (Figure 3-13). The yeast would use this reaction to furnish sulfite for reduction to sulfide for cysteine, serine and methionine synthesis. Normally this occurs only under conditions of insufficient amino acids in the juice. Wines with low amino acids develop a slight H_2S aroma (Bell et al., 1979).

Monk (1986) points out that the H_2S formed in wine can come

FIGURE 3-13. Details of the sulfur pathways leading to formation of sulfite, sulfide, and sulfate in yeast. Permeases allow for transport into or out of the yeast. 1 Sulfate Permease; 2 ATP-Sulfurase; 3 APS-Kinase; 4 PAPS-Reductase; 5 Sulfite Permease; 6 Sulfite Reductase; 7 Serine Sulfhydrase; 8 OAS-Sulfhydrase; 9 OAHS-Sulfhydrase; 10 Methyl Transferase; [APS = Adenosine-5-Phosphosulfate; PAPS = 3'-Phosphoadenosine-5'-Phosphosulfate; OAS = o-Acetylserine; OAHS = o-Acetylhomoserine.]

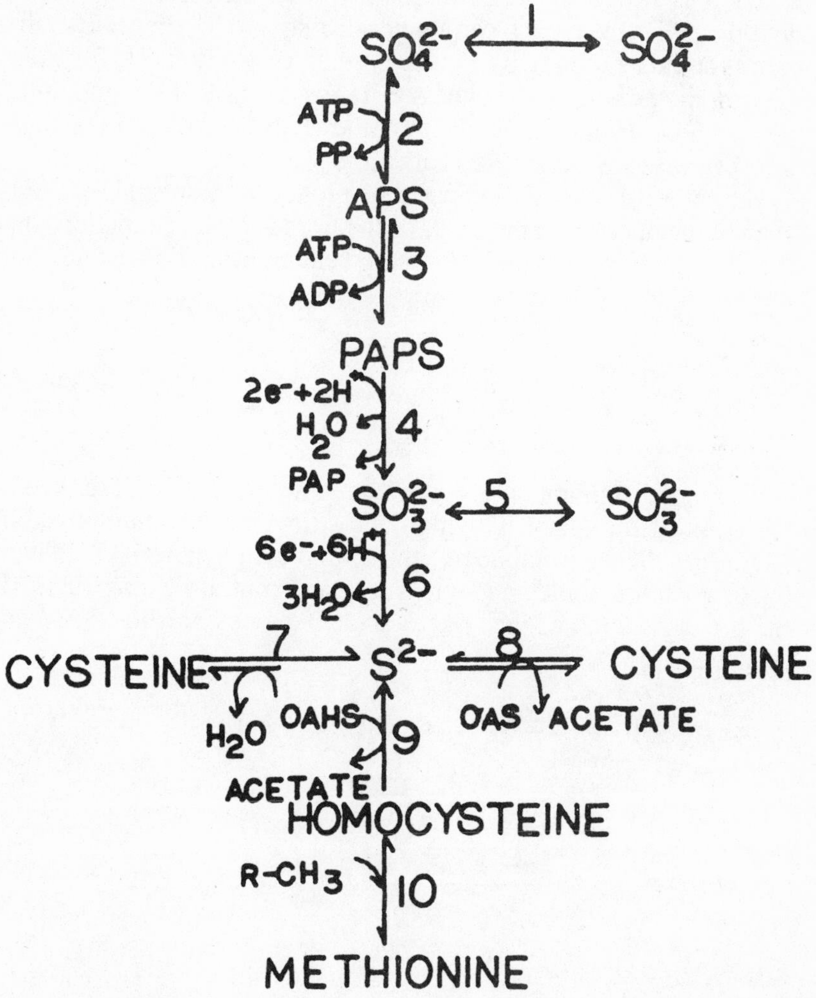

from numerous sources. One of the main sources, besides elemental sulfur, is by a deficiency of amino acids and pantothenate. These are necessary for serine and methionine production. When the vitamin is scarce, H_2S is made in excess and excreted. This can be avoided by the addition of ammonia to stimulate yeast growth and biosynthesis.

There is also an active permease that can transport sulfite directly into the cell. One reason certain yeasts may produce more H_2S than others could be overactive sulfite permease. Montrachet, for example, produces more H_2S when sulfite is present that do some other yeasts. Only about the same amount of H_2S is produced compared to other yeasts if no sulfite is added to the juice.

Higher alcohols (fusel oils) can be produced by several reactions. If an abundance of amino acid, or the need for transamination, then the Ehrlich-Neubauer and Fromberg scheme would be active.

$$R_1\text{-Amino acid} + R_2\text{-}\alpha\text{-Keto acid} \xrightarrow{\text{transaminase}} R_1\text{-}\alpha\text{-Keto acid} + R_2\text{-Amino acid}$$

$$R_1\text{-}\alpha\text{-Keto acid} \xrightarrow{\text{decarboxylase}} R_1\text{-Aldehyde} \xrightarrow[\text{NADH} \curvearrowright \text{NAD}^+]{\text{dehydrogenase}} R_1\text{-Alcohol}$$

The most common pathway in wine fermentation is described by the work of Ingraham, Guymon and Crowell, and summarized by Guymon (1966). In this case, the amino acid synthesis is actively making carbon skeletons from sugar and transaminating them. In the process, some of the α-keto acids become decarboxylated and reduced.

$$\text{R-COCOOH} \xrightarrow[\text{Acetyl CoA}]{\text{acetoransferase}} \text{R-COH(COOH)-CH}_2\text{COOH} \xrightarrow{\text{rearrangement}}$$

$$\text{R—CH(COOH)—CH(OH)—COOH} \xrightarrow[\text{NAD}^+ \curvearrowright \text{NADH}]{\text{dehydrogenase}} \text{R—CH(COOH)—C(=O)—COOH}$$

$$\xrightarrow{\text{decarboxylase}} \text{R—CH}_2\text{—C(=O)—COOH} + CO_2$$

The third method involves the condensation of an α-keto acid with either pyruvate or active acetaldehyde (TTP), then rearrangement, dehydrogenation, removal of water, and then as above, the decarboxylation and reduction to the alcohol.

$$R_1\text{—}\overset{\overset{\large O}{\|}}{C}\text{—COOH} \xrightarrow[\text{TPP}]{\text{condensation}} CH_3\text{—}\overset{\overset{\large O}{\|}}{C}\text{—}\underset{\underset{\large COOH}{|}}{\overset{\overset{\large R_1}{|}}{C}}\text{—OH} \xrightarrow{\text{rearrangement}}$$

$$R_1\text{—}\underset{\underset{\large OH}{|}}{\overset{\overset{\large CH_3}{|}}{C}}\text{—}\overset{\overset{\large O}{\|}}{C}\text{—COOH} \xrightarrow[\text{NADH NAD}^+]{\text{dehydrogenase}} R_1\text{—}\underset{\underset{\large OH}{|}}{\overset{\overset{\large CH_3}{|}}{C}}\text{—}\underset{\underset{\large H}{|}}{\overset{\overset{\large OH}{|}}{C}}\text{—COOH} \xrightarrow{\text{hydrolase}}$$

$$R_1\text{—}\underset{\underset{\large H}{|}}{\overset{\overset{\large CH_3}{|}}{C}}\text{—}\overset{\overset{\large O}{\|}}{C}\text{—COOH}$$

n-Propanol is formed from either the first two pathways. Isobutanol and active amyl alcohol are formed from the first or third pathway. Isoamyl alcohol is formed from all three. The amounts of fusel oils that are found in wine vary with yeast strains, with fermentation temperature and nutrient level. Excess solids during fermentation or aeration both caused increased fusel oil formation. Efforts to develop yeasts that make lesser amounts of fusel oils have only been partially successful.

The amino acids are transported into the yeast cells by several permeases. Once in the cell, they can be used directly for synthesis of protein and other compounds. Also, they can be transaminated and the amine group exchanged with an α-keto acid to form a desired amino acid. α-Ketoglutaric acid is a general receptor for the amino group and glutamic acid the donor for other transaminations. The α-ketoglutaric acid is a product of the citric acid cycle. The actual synthesis of the amino acids can be divided into several groups. The amino acids in each group have a common starting point.

Amino Acid Synthesis Families

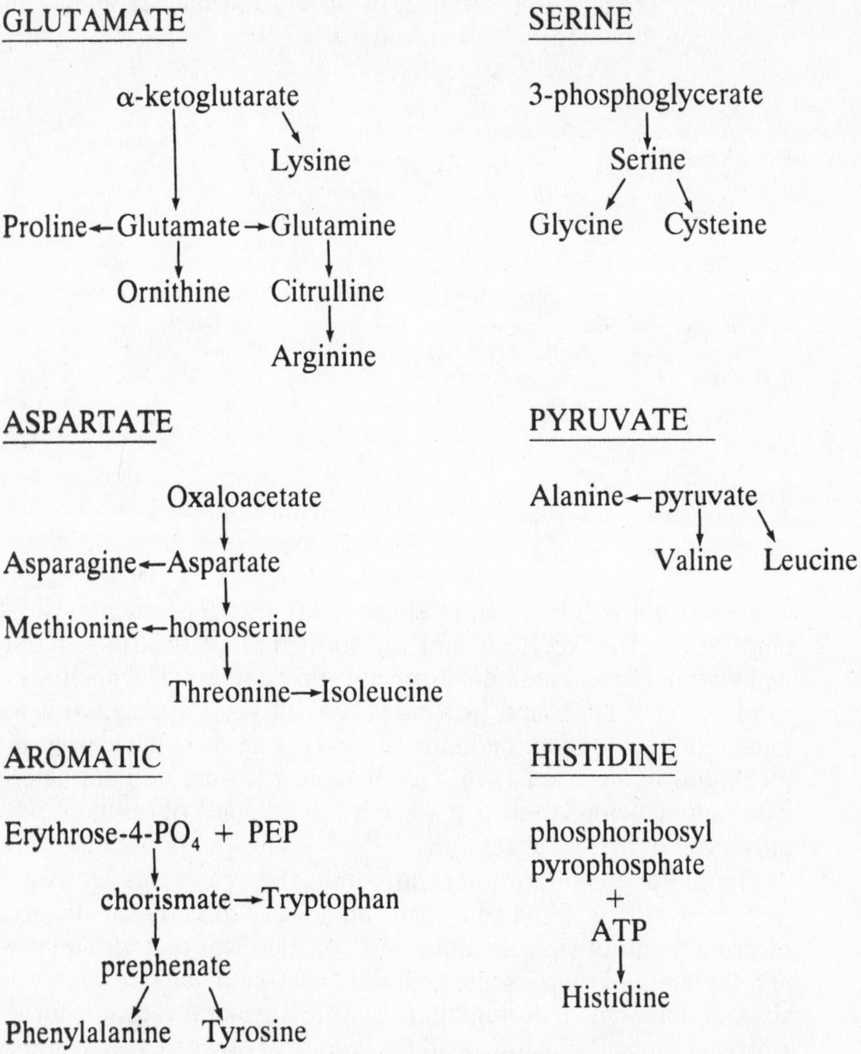

GLUTAMATE

α-ketoglutarate → Lysine
α-ketoglutarate → Glutamate
Proline ← Glutamate → Glutamine
Glutamate → Ornithine
Glutamine → Citrulline → Arginine

SERINE

3-phosphoglycerate → Serine
Serine → Glycine, Cysteine

ASPARTATE

Oxaloacetate → Aspartate
Asparagine ← Aspartate
Aspartate → homoserine
Methionine ← homoserine
homoserine → Threonine → Isoleucine

PYRUVATE

Alanine ← pyruvate
pyruvate → Valine, Leucine

AROMATIC

Erythrose-4-PO_4 + PEP → chorismate
chorismate → Tryptophan
chorismate → prephenate
prephenate → Phenylalanine, Tyrosine

HISTIDINE

phosphoribosyl pyrophosphate + ATP → Histidine

The catabolism of the amino acids is of interest in several reactions. Especially the reactions involved with arginine. The yeast, by the action of arginase, form ornithine and urea. The urea is broken down by a two-step enzyme complex into ammonia and CO_2. The ornithine can be broken down further into proline by a three-step reaction. In the process, the ornithine terminal amino group is transaminated onto α-ketoglutaric acid to form glutamic acid. Thus, three moles of ammonia can be recovered from one mole arginine.

This reaction is interesting for other reasons also. The small amount of urea that is in the cytosol of the cell can either be transported by known permease or diffuse into the wine. Urea in any significant amounts can, depending on time, temperature, pH and ethanol concentration, be transformed into ethyl carbamate.

$$NH_2-C(=O)-NH_2 + C_2H_5\text{-}OH \rightarrow C_2H_5\text{-}O-C(=O)-NH_2 + NH_3$$

Of further interest is the reaction for the catabolism of proline to glutamic acid. This occurs only under oxidative conditions and when the other amino acids and ammonia have been used. Figure 3-14 shows the fermentation of a juice with less than necessary amounts of amino acids present, and the effect of adding arginine to the medium. These fermentations were done in small containers and aeration occurred when sampling was done.

Acetoin, diacetyl and 2,3-butanediols are formed via either pyruvate or acetaldehyde (Figure 3-15). The corresponding five-carbon analogs can be formed in a similar manner but to a lesser extent. The 2,3-butanediols and propanediols are formed by further reduction of acetoin and its propane analog by the proper dehydrogenases. Acetoin and 2,3-butanediols do not have any significant sensory influence. Diacetyl, formed mainly during malolactic activity of certain bacteria, can cause rancid butter-like off-aroma if too much is formed. At threshold levels, it is less unpleasant.

The formation of acetoin via pyruvate by pyruvate carboxylase to acetyl-CoA is activated by excess ATP. This leads to the pathway of the citric acid cycle to produce α-ketoglutarate and finally succinic acid. There are three NADH produced in the formation of succinic acid. Oura (1977) proposes this takes place to maintain the proper reducing conditions within the yeast. Another step involving

FIGURE 3-14. A Chardonnay juice from a poorly fertilized vineyard at 50 mg/L arginine and with additions of arginine. Fermented with *S. cerevisiae* Montrachet at 20° C with slight oxidation during fermentation. Arginine •, and proline ∘.

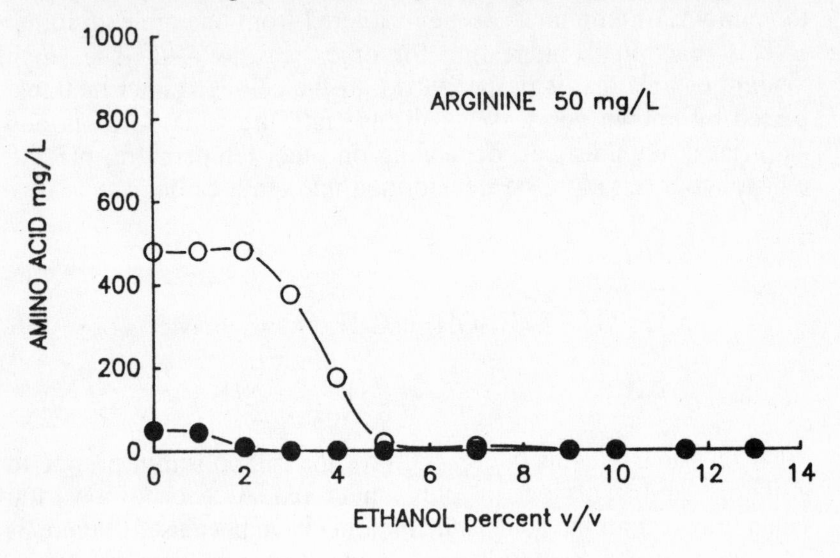

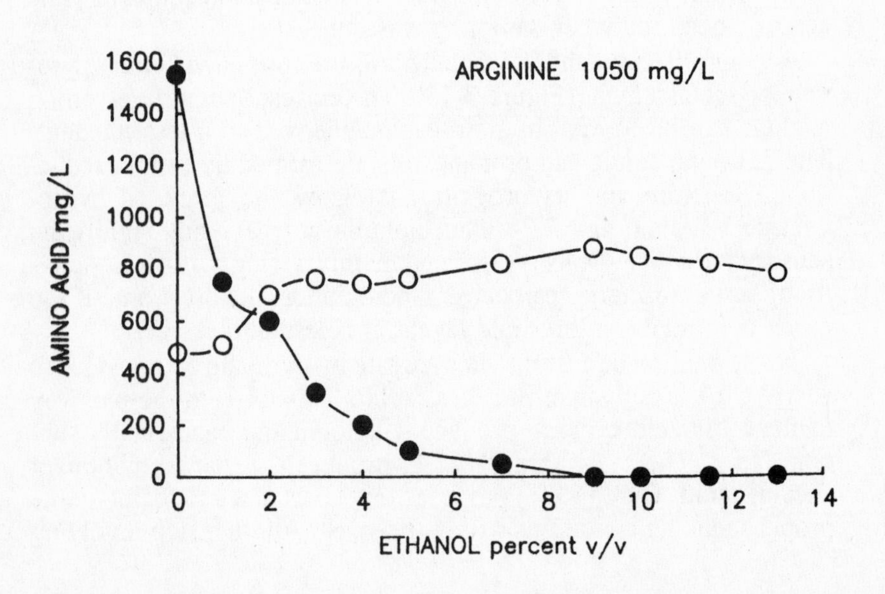

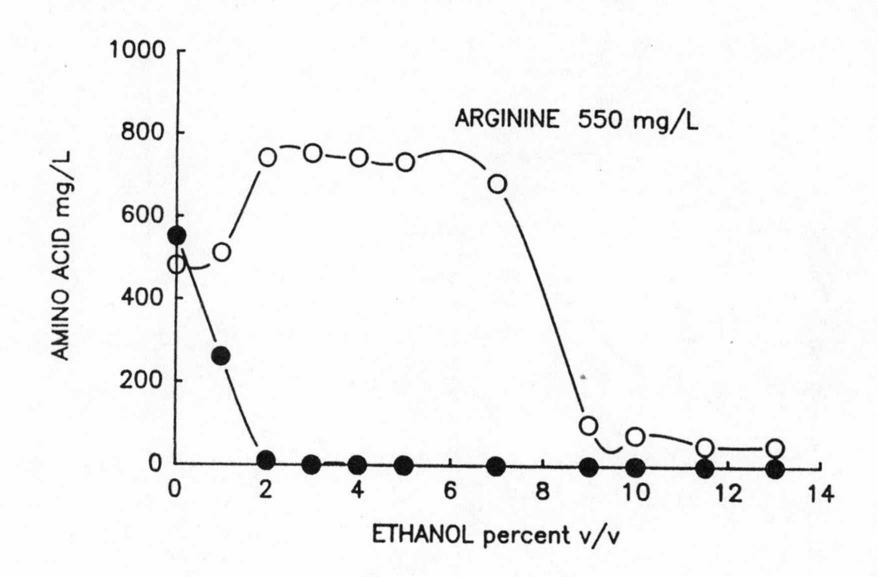
ARGININE 550 mg/L
AMINO ACID mg/L
ETHANOL percent v/v
1000
800
600
400
200
0
0
2
4
6
8
10
12
14

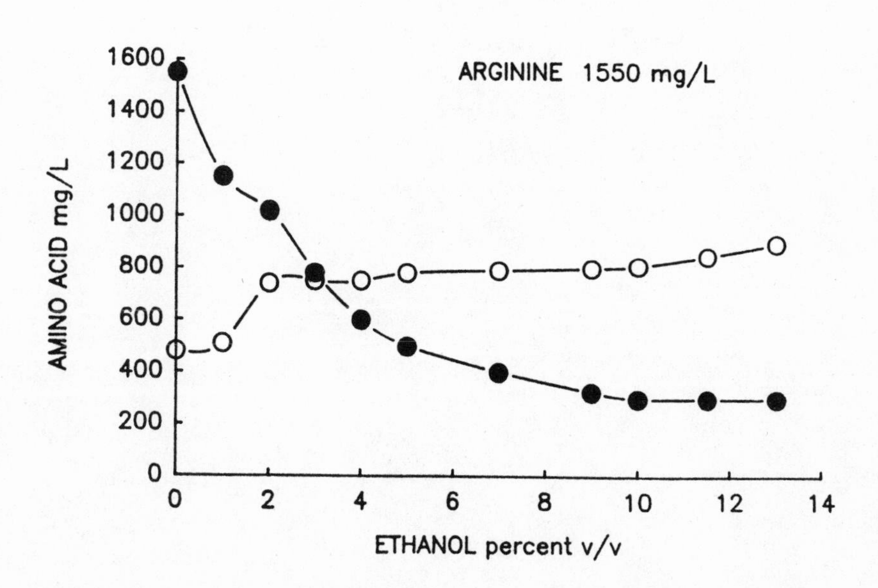
ARGININE 1550 mg/L
AMINO ACID mg/L
ETHANOL percent v/v
1600
1400
1200
1000
800
600
400
200
0
0
2
4
6
8
10
12
14

FIGURE 3-15. Pathways for the formation of acetoin and 2,3-butanediol.

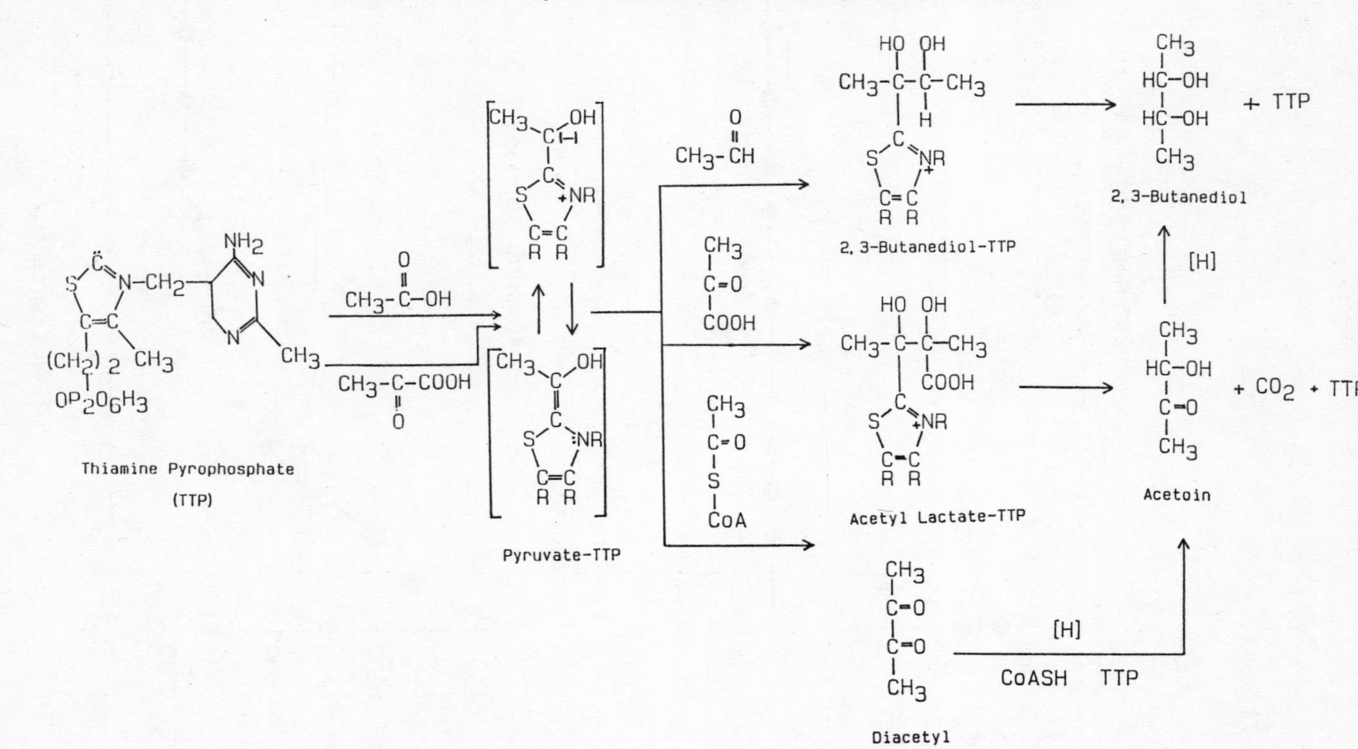

pyruvate is the reaction of pyruvate with acetyl-CoA to form acetoin. Further, the acetoin is reduced to 2,3-butanediol, using up to one NADH in the process. Two isomers of 2,3-butanediol are produced, meso and D(-) or levo. The other route for acetoin formation is via acetaldehyde activated with thiamine pyrophosphate. This route is likely in the formation of acetoin under aeration conditions such as flor sherry production. No 2,3-butanediols form even though acetoin is formed in abundance (Ough and Amerine, 1972).

WINE COMPOSITION

The composition of wine depends on many factors. The grape composition is the primary determinant; but handling, fermentation and cellar treatments, as well as storage and transport conditions beyond the winemakers' power to control are factors. The political aspects of wine import-export have caused some rather intricate methods of analysis to try to avoid sophisticated wines being imported. Sometimes rules and analysis are used solely for protectionism with no rational scientific reason. Other times, for health reasons or plain fraud, careful consideration is given to the composition by government authorities. There are efforts on the part of the industry institutes to self-regulate or to have quasi-legal groups control the wine by analytical limits. For example, the definition of sherries in California is based on the sugar content for cocktail, dry and sweet. Not everyone abides completely, but most do. On the other hand, varietal wines must contain 75% of the grape of that cultivar. This is controlled by weight tags and records of yield and blending. There is no way, at present, it could be done by chemical analysis. On the other hand, part per billion of some compounds can routinely be detected in wines.

Alcohols

Ethyl alcohol is closely regulated in the USA for tax purposes. The amount of tax is dependent on the alcohol content. Seven percent or over to 14% v/v is taxed as table wine at $1.07/wine gallon. Over 14.0% v/v to 21% v/v, the tax is $1.57/wine gallon and from 21.0% v/v to 24%, it is $3.15/wine gallon. Champagne is taxed

$3.40/wine gallon and carbonated wines at $3.30/wine gallon. Over 24% v/v the tax is $13.50/proof gallon.

All wines fall within the designated ranges except some sweet table wines and late harvest wines, which are over the 14% v/v but less than the dessert and sherry minimum limits of 18 and 15% v/v, respectively.

Most table wines are between 10-14% v/v. Only special wines and coolers, and such, fall in the 7-10% v/v area. Major alcohols, other than ethanol, are summarized in Table 3-4. Trace amounts of terpenols and many other alcohols have been found and identified.

Carbonyls

Carbonyl compounds, except for acetaldehyde, acetoin, and diacetyl, are the only ones found in trace amounts in table wine. In dessert wine, sherries and concentrate, hydroxymethyl furfural (HMF) can be at fairly significant levels. In Tokay wines, values can also be high. This indicates some heating of the grapes or wine or the use of concentrate. HMF forms from fructose only by heat application and is not normally found in other wines at these levels. In most table wines, the average values for HMF are less than 3 mg/L. Table 3-5 gives a tabulation of some carbonyl values.

Esters

Most of the volatile esters are formed during fermentation as explained earlier. However, the esters of the non-volatile organic acids form slowly with time. As with the volatile esters, the rates of change to equilibrium are slow and temperature dependent. Given in Table 3-6 are the values for many of the volatile and non-volatile esters that have been measured. There are numerous other esters found in wines. Most are only in very small quantities and have not been adequately quantified and are probably not too significant (Baumes et al., 1986).

Nitrogen Compounds

Nitrogen values in wine depend on the amount left in the wine at the end of the fermentation. This can be further altered up or down

Table 3-4. Values of Various Alcohols Measured in Table Wines.

Alcohol mg/L	Number of samples	Range	Average
Methyl			
White table	385	0-240	78.8
Red table	389	0-269	129.8
1-Propyl	527	7-68	27.6
Isobutyl	724	7-174	54.6
Active amyl	269	19-96	52.3
Isoamyl	321	83-352	201.6
1-Hexanol	79	1.3-12.0	4.0
2-Phenethyl	79	5-74	30.0
Glycerol g/L	651	1.11-23.0	7.33
2,3-Butanediols	314	16-1615	522.5
Sorbitol	589	5-394	70.0
Mannitol	72	84-1401	468.5
Erythitol	44	33-272	111.6
Arabitol	44	13-329	89.8

Summarized from various sources.

by finings done later to the wines. Bentonite will decrease the nitrogen, while materials such as casein and other proteins can be partially solubilized and retained. Red wines, because of the higher fermentation temperatures, tend to use up the nitrogen more completely than white wines fermented at lower temperatures. Some values for nitrogen components are shown in Table 3-7. For a compilation of volatile nitrogen compounds in wine and other fermented products, see Ough (1984).

Table 3-5. Values for Carbonyls found in Table Wines.

Carbonyls mg/L	Numbers of samples	Range	Average
Acetaldehyde	861	3-491	55.2
Acetoin	321	0.7-138	10.3
Diacetyl	675	0.1-7.5	1.97

Table 3-6. Values for Esters Found in Wines.

Ethyl mg/L	Number of samples	Range	Average
Ethyl acetate	828	11-232	56.40
n-Propyl acetate	14	0.04-0.8	0.29
Isobutyl acetate	18	0.00-0.5	0.23
Isoamyl acetate	56	tr-9.3	2.20
n-Hexyl acetate	43	0-1.0	0.20
Phenethyl acetate	44	0-1.14	0.22
Ethyl butyrate	7	0.2-0.44	0.33
Ethyl caproate	44	tr-1.8	0.84
Ethyl caprylate	44	tr-2.1	1.15
Ethyl caprate	44	tr-0.9	0.45
Ethyl succinate	16	0.2-6.3	0.76
Mono-caffeoyl tartrate	10	70.9-233.8	120.70
Mono-p-coumaroyl tartrate	10	8.3-33.8	18.10
Mono-feruloyl tartrate	10	1.6-15.9	5.6
Methyl anthranilate	30	0.14-3.50	0.85

Table 3-7. Values of Some Nitrogen Components in Table Wine.

Nitrogen component	Number of samples	Range	Average
Total nitrogen as mg/L N	737	70-980	320.6
α-Amino nitrogen[1] as mg/L N	317	3-452	74.0
Amino acids as mg/L amino acids[2]	147	373-4224	1947.0
Aspartic acid	147	0-60	11.2
Threonine	147	0-54	5.0
Serine	147	0-106	7.0
Asparagine + glutamic acid	147	0-187	39.9
Glutamine	147	0-291	38.2
Proline	147	317-3830	1463.4
Alanine	147	0-314	58.4
Citrulline	147	0-56	4.9
Valine	147	0-73	7.8
Methionine	147	0-23	5.2
Isoleucine	147	0-27	3.5
Leucine	147	0-39	3.0
γ-Aminobutyric acid	146	0-155	57.5
Tyrosine	147	0-34	0.3
Phenylalanine	147	0-50	4.8
Ammonia	147	0-29	3.2
Histadine	147	0-50	10.4
Ornithine	146	0-130	19.5
Lysine	146	0-46	8.7
Arginine	147	0-1372	191.6
Biogenic amines mg/L			
Histamine	838	0-49.1	2.47
Tyramine	462	0-12.2	1.00
Nitrate mg/L	1331	0-53.7	7.75

[1]Formol titration or cation exchange and ninhydrin.
[2]By amino acid analyser, ninhydrin post column, including proline, all values summed. All from red and white varieties from Napa Valley.

Phenol Compounds

Phenol components of the grape determine color of the wine and contribute to its flavor and aroma. Total phenol content of the white table wines vary from around 50 to 1000 mg/L and average about 200 mg/L. The Folin-Ciocalteau method measures the phenols as gallic acid equivalents (GAE). Red wines can have from 900 to 2500 mg/L as GAE and probably average close to 1200 mg/L GAE. The color pigment of the grape and new wines are the anthocyanins. These are made up of delphinidin-, cyanidin-, petunidin-, peonidin- and malvidin-3-mono-glucosides and the acetate and coumarate acylated compounds.

For young red wines, the anthocyanins can be from 200 to 500 mg/L. There is significant variation between varieties. In non-*vinifera Vitis* sp., diglucosides are present. Measurement of malvidin diglucoside is a criteria for determining the purity of the *V. vinifera* wine. If over 5% in diglucosides, non-*vinifera* blends can be detected. Catechins are associated with the skins and seeds of the grape. For white wines, the catechin values are probably around 10-20 mg/L. Flavonols and flavonones are found in very insignificant amounts in white wines. Red wines can have 20-100 mg/L of flavonols. Quercetin-3-O-glucuronoside has been reported as a normal constituent of wines. It arises from the grape. The corresponding glycosides are not readily found, as they hydrolyse during fermentation according to Alenso (1986). Non-flavonoid volatile phenols are present individually at less than 0.1 mg/L except the amino acid-related tyrosol, tyrosol acetate and tryptophol. These are found from 0.1 to 5.0 mg/L in wine. The volatile non-flavonoid phenols—salicylic, vanillic, gentisic, syringic, p-coumaric, gallic, ferrulic and caffeic acids—range from 2.0 to 16 mg/L in wine. For a complete composition summary of the phenols in wines, see Singleton (1988).

Nonorganic Compounds

The nonorganic constituents of wine consist of many anions and cations. The cation content is mainly potassium, sodium, calcium and magnesium. Iron and copper are of interest because of their importance to the stability of the wine. Lead has been of interest from possible contamination from old foils. Zinc has also been measured with some consistency. Table 3-8 summarizes the values for these and other trace cations reported in wines.

Traces of most other metals were detectable at very low levels. According to Marlatta et al. (1986), grapes within 50m or closer to

Table 3-8. Values of Various Cations Found in Wines.

Cation mg/L	Number of samples	Range	Average
Potassium	754	90-2040	833.00
Sodium	754	3-320	40.40
Calcium	549	6-310	90.60
Magnesium	459	21-245	107.20
Iron	448	0.3-22	4.20
Copper	373	tr-2.4	0.17
Lead	1619	0-1.26	0.13
Zinc	1077	tr-11.7	1.17
Aluminum	77	0.67-5.4	1.69
Arsenic	110	0.001-0.53	0.093
Antimony	75	0.0012-0.009	0.0037
Barium	65	0.02-0.22	0.12
Boron	121	3.9-80.3	17.80
Cadmium	180	tr-0.049	0.004
Cesium	74	0.0002-0.0047	0.0017
Chromium	213	0.004-0.81	0.070
Lithium	65	0.005-0.11	0.018
Manganese	464	0.09-17.4	1.36
Mercury	63	0.02-0.65	0.06
Molybdenum	51	0.01-0.33	0.05
Nickel	59	0.01-0.40	0.05
Rubidium	131	0.58-4.99	1.70
Silicon	51	12.6-28.9	19.60
Strontium	88	0.14-3.13	1.01

Various sources.

a highway can have increasing levels of cadmium, as well as lead, due to exhaust fumes.

Inorganic anions are mainly chlorides, phosphates and sulfates (Ough and Amerine, 1988). Ranges of these compounds are from 5-200, 25-850 and 70-3000 mg/L, respectively. The averages are

50, 300 and 775 mg/L, respectively, for an extensive number of analyses. Bromides from 257 analyses ranged 0-0.54 mg/L with an average of 0.10 mg/L. For 130 samples of wines, iodides ranged 0.1-0.4 mg/L and an average value of 0.15 mg/L. Fluoride in 845 samples was found from tr-0.95 mg/L. Only a few values over 0.5 mg/L were reported (less than 4%).

Slight increases in phosphate can occur from the use of diammonium phosphate or improper rinsing of phosphate cleaning agents. Sulfates can result from sulfur dioxide oxidation or plastering of wines to reduce pH.

There is very little concern for the halides unless illegal additives are used. Fluorides can be increased slightly if water in the winery was fluoridated or certain pesticide residues are present. Fluoride can inhibit yeast growth and fermentation if in excess of a few mg/L in the juice.

Organic Acids

The organic anions are tartrates, malates, succinates, lactates and citrates. They make up the bulk of the anions in particular as tartrates and malates. The relative concentrations of the tartrate and malate vary with the climate and maturity and the total amounts vary with the crop load and climate.

Tartrates as tartaric acid very from about 2 g/L to 8 g/L. The malates as malic acid can be from traces to about 5 g/L or slightly higher in cool climate grapes or immature grapes. Citrates as citric acid usually are less than 1 g/L, most of this is added before bottling for iron stabilization. Succinate can vary with yeast strain. The reported levels are from 0.5 to 7.5 g/L for table wine. L(+) lactic acid formed from malolactic fermentation varies from 3 to 6 g/L. Natural D(-) lactic acid is present from 0.1 to 1.0 g/L.

Other Compounds

Many important constituents in wine remain to be quantified and investigated. Measurement of the various non-volatile materials, both simple and complex, need further work. It has only been recently that urea was quantified in wine (0.1 to 100 mg/L). The various protein-phenol complexes which are soluble need much investigation. Many trace nitrogen compounds have never been quantified. These include peptides as well as other yeast products. Lipids are just now coming under more intense investigation in wines. As the equipment to do these measurements is developed, so will the knowledge of the composition of the wine.

Baumes et al. (1986) determined the volatile constituents in the wine from five red and five white cultivars. They positively identified 122 compounds, which may or may not have contributed to the odor, and discussed the source of some of the volatiles. The complexity of this type of analyses restricts this kind of research to universities or research institutes. Many of the aroma-producing compounds are there in μg/L levels, or less, and are extremely difficult to quantitate.

Chapter 4

Clarification and Fining of Wines

Wines are freed of sediment and yeast after the main fermentation ceases. In special cases where yeast contact is desired for extraction of flavors and aromas, then the racking at the end of fermentation is delayed. In some instances, the leaving of the wines on the lees to benefit malolactic fermentation is desirable. Also, in some situations red table wines are left on the skins for further extraction of flavor and color. Undue delay in clarification can lead to spoilage and off-flavors if care is not taken.

Wine will expand when it warms up. If the wines have been kept at 10°C and warm up to 20°C, about 0.4% expansion can occur. This can be a rather large amount in a big tank. For 50,000 gal the amount is 200 gal, a significant amount if lost. Leave extra space and top off after the wine comes to temperature.

WHITE WINES – FINING

White wines can be clarified rapidly by centrifugation or by fining with bentonite with later filtration. Just rackings alone can cause most wines to clarify after several rackings over time. The bentonite treatment not only adsorbs and settles out colloidal and yeast material but also removes heat unstable proteins.

Delay in initially taking care of wine as it finishes fermentation can allow oxidation and growth of *Mycoderma* (*Candida* sp.) or *Acetobacter*. Surface yeast such as flor yeast can grow and spoil the wine. To prevent this from happening, oxygen must be excluded from the surface of the wine. Racking the wine into full tanks with a minimum of head space will prevent microbiological spoilage. That space can be filled with CO_2 to further protect the wine from contamination and oxidation. At this point the wine should be dry or have reached a desired sugar level.

White wines fermented in barrels can be easier to clarify because of the small volume and the adsorption of the colloidal material to the wood.

Bentonite

Bentonite (montmorillonite, a term for certain aluminum-silica clays) is a sodium-dominated form of the clay. Its use goes back to when Saywell (1934) first used bentonite in California on vinegar as a clarifying agent. It forms a milky, creamy mixture with water, usually used in about 5% wt/vol mixture. The adsorption of proteins from wine is more effective at lower pH values. The increased positive charges on the proteins react with the negatively charged bentonite. Other types of clays and bentonite are usually less effective than the Wyoming bentonite.

Bentonite can be prepared by slowly mixing the bentonite into sufficient water to make a thoroughly blended cream. More than enough bentonite should be weighed originally to fine the wine in question. The creamy water mixture should be allowed to stand 24 hrs and mixed with a blender again. A minimum of water should be used. After a further 24 hrs of standing to allow the bentonite to swell, the mixture is brought to a volume (usually a 5% solution wt/vol).

A test should be done to determine the amount of bentonite required. The wine must not be fermenting. Settle overnight at the temperature the wine is to be stored. Examine the samples and determine which level gave the best clarification results. Over-fining can cause difficulty in further clarification. This test can be done in 750 ml wine bottles.

If a 5% wt/vol bentonite preparation was prepared (for ease of calculation, use the metric system for the test), then each ml of a 5% slurry is equivalent to 0.05 g (50 mg). Calculate the amount needed to be added for 750 ml of the wine to bring wine to 100 mg/L of bentonite.

$$100 \text{ mg/L of bentonite} = \frac{100 \text{ mg/L} \times 750 \text{ ml}}{50 \text{ mg/ml} \times 1000 \text{ ml}} = 1.50 \text{ ml of the slurry}$$

Add amounts of the slurry to give 100, 150, 200, 250 and 300 mg/L of bentonite.

The amount as lbs/1000 gal can be determined by knowing that 1 lb/1000 gal is equal to 120 mg/L. So if one tested for addition of 1 lb/1000 gal you would add 120/100 × 1.50 or 1.80 ml/L.

Bentonite reacts very rapidly to adsorb the proteins and other materials. Blade and Boulton (1988) found 30 sec was all the time required for complete protein adsorption in a model wine solution. They also found that the rate of adsorption was independent of temperature but did depend on pH and ethanol content.

Sodium bentonite has a better adsorptive capacity for protein than the calcium form. To get good mixing, the bentonite should be proportioned into a transfer line as the wine is racked. Bentonite can remain in the wine for some time (weeks) without adverse sensory effects.

Certain grape varieties have more protein than others. Sauvignon blanc and muscat varieties can be particularly high in amount.

Postel et al. (1986) examined the metal changes in wines treated with bentonite. Aluminum increased in the wine depending on whether its was Na or Ca bentonite. The latter yielded more Al. Zn, Cu, Sn and Cd did not increase, Cr increased slightly, as did Pb. Pb never increased over the German legal limit of 0.3 mg/L. K was slightly decreased.

Silica Gel and Gelatin

Silicic acid in a colloidal suspension is used with gelatin for wine clarification and removal of proteins and tannins. If the pretreatment is to remove tannins, the gelatin (bloom #80-130) is added first with thorough mixing. Then the silicic acid is added, mixed and the wine settled and filtered. When the prime purpose is to remove protein, the silica gel (with or without bentonite) is added first. Then the gelatin is added. Bearzatto (1986) gives some levels of use in Table 4-1. These are median levels. If more or less is required, maintain the silica gel-gelatin ratio.

Table 4-1. Silica Gel-Gelatin Ratios for Fining.

Silica gel[1]		Gelatin		
Per hectoliter	Per 1000 gal	Per hectoliter	Per 1000 gal	Case
25 ml	946 ml	1.25 g	47.3 g	Normal wines
50 ml	1892 ml	2.50 g	94.6 g	Easy to clarify, but young wines with much cloud
100 ml	3785 ml	10.00 g	378.5 g	Difficult to clarify
50 ml	1892 ml	5.00 g or 10.00 g	189.2 g or 378.5 g	High tannin content

[1]30% solution of silicic acid.

The effectiveness of the silica gel-gelatin ratio must be tested by trials with the wine in question. The values in Table 4-1 are guide amounts only. Some silica gels are less effective than others. The gelatin is dissolved in water (one part gelatin and five parts H_2O). The water may be warmed to get the gelatin in solution but should not be boiling.

Sparkaloid

Sparkaloid® is a proprietary compound of several materials sold ready for use. It can be useful for wines that are difficult to clarify and filter.

Casein

Press wines or wines with excess phenols are fined by the addition of casein (milk protein) and tannin. The levels of the materials may vary slightly with the type of wine. In Chapter 5 the use of this agent for white wine color removal and color stabilization and other purposes is discussed.

Isinglass

Isinglass is the dried air bladder of the sturgeon. Dissolve 1/2 kilogram (about 1.1 lbs) of isinglass flakes in 50 liters (13.2 gal) of water containing 50 g of tartaric acid and 20 g of potassium bisulfite. It should be stirred often for the first few hours. After several days all the lumps are mashed with a brush. The material is strained through a fine mesh. The amount used is around 250 ml to 500 ml per 50 gal. The rate of application suggested by Peynaud (1984) is 1 to 2.5 g/hl. He notes it should only be used for white wines. It has been used for red wines in California at times. Isinglass has a drawback of having a bulky precipitate that is glue-like and difficult to remove.

Polyvinylpolypyrrolidone

Polyvinylpolypyrrolidone (PVPP) is used for phenol removal from white wines. From 30 to 500 mg/L, as determined by trial fining, can be used to reduce the astringency of wine. GAF™ has test tablets, Unitest 50®, that can be added to the wine in question to determine the amount to use. The compound, a powder, is added to the tank during mixing. It is stirred for one hour, settled and the wine racked or filtered off the sediment. More recently, filter sheets have been impregnated with the PVPP and the wine filtered through the sheets. Up to 50% removal of the phenols in the first 3 hrs and about 30% removal after 7 hrs has occurred. These sheets can be reused by backwashing with warm water, then 0.5% NaOH and then neutralization with 0.5% H_2SO_4 or H_3PO_4 and a final hot water rinse. They can be reused up to 10 times. PVPP is insoluble in wine. The level that is permitted is 1 mg/L in the final product.

RED WINES – FINING

Fining of red wines is done primarily to reduce the astringency of the wine. Over the years, several proteins have been used for this purpose – among them blood, egg whites, gelatin, and isinglass. Recently, blood has not been popular for esthetic reasons. Dried blood was found in recent tests in the author's laboratory to be as effective as gelatin for removal of phenols. Usually dried egg whites are used rather than fresh egg whites. Gelatin comes now from vegetable sources rather than animal, as previously used.

Gelatin comes in two types, one is prepared using acid and is positively charged. The other is prepared by a base treatment and is less positively charged. Since the phenols which are to be fined are negatively charged, the acid-prepared gelatin is the choice. Higher bloom number gelatin is more effective in removing larger molecules but the low bloom number gelatin is much cheaper.

Egg whites are separated and solubilized in cool water. Enough salt (NaCl) is added to clarify the mixture. The equivalent of 5-10 eggs may be required to fine 50 gals of red wine. If dried egg whites are used, it takes about 8-16 g/50 gal. Egg protein is only used for red wines.

All of these protein materials act the same. They react with phenolic compounds to remove them from the wine. While there may be some minor differences in which phenols are removed, that is a small difference. Isinglass can change the flavor of the wine slightly and perhaps give a style to the wine. Gelatin, egg whites and dried blood do not cause any major flavor or aroma changes in the wine. They can reduce astringency and make the wine taste smoother and less tannic.

As before, test wine trials need to be made to determine how much fining agent is to be added. All of these materials are soluble, to a degree, in wine. Excessive additions can give problems with clarity of the wine as well as add to the expense.

The compounds which are most susceptible to removal by protein fining are the complex phenols or flavonoids that will or have polymerized. These are also related to the astringency feel. Some winemakers fine a red wine one or more times during its aging. Part of the reason for this is to allow the polymers to form. Several small additions over a time are more effective than a single application. Table 4-2 gives the ranges of the various fining agents used in red table wines and comments on effectiveness. It is assumed that a fining agent does not remain in a wine. This is not so for soluble proteins. Table 4-3 gives data from Watts et al. (1981) showing that a fair percentage can remain in the wine. These tests concerned materials causing allergies.

Tests have been done with wine on persons allergic to some of

Table 4-2. Amounts of Protein Fining Agents Used for Red Table Wines.

Agent	Range used as mg/L	Comments
Gelatin	50-200	Remove a portion of the color as do all these agents.
Egg white	50-120	Several small finings over the aging period are probably superior to one large fining.
Isinglass	10-30	Changes the flavor of red wines, use carefully.
Blood	50-250	Dried blood reacts similarly to gelatin.

Table 4-3. Amounts of Fining Agent Remaining in a Wine after Treatment, Settling, Racking and Filtration.

Fining Agent	Levels added mg/L	Remaining in finished wine[1] mg/L
Casein (with tannic acid)	240	32.60
(white wine)	120	21.20
	60	12.70
Egg whites	240	2.32
(red wine)	120	1.34
	60	1.04
Isinglass	40	1.52
(red wine)	20	0.39
	10	0.30
Pectic enzyme[2]		
(treatment of the white juice)	0.40	0.45

[1]Average of 3 wines.

[2]Pectinex® 3X at 20 ml/ton grapes.

these proteins. Test results indicated that, in certain instances, ingestion of the protein-fined wines caused minor positive allergic response from highly allergic persons (Marinkovich, 1983). These individuals are very few and the response was not life-threatening. Residual gelatin was not measured. There has never been a report of any person being allergic to gelatin.

It is easily possible to follow the changes in the total phenols and color. The usual Folin-Ciocalteau method and optical density readings at 420 and 520 nm on the spectrophotometer are adequate. Figure 4-1 gives an example of how the changes occur in a wine fined with gelatin and with dried blood at various levels.

Fining of red wines requires some experience to know what to

FIGURE 4-1. The effect on the total phenol and the color for a red wine fined with dried beef blood ∘ and with gelatin • .

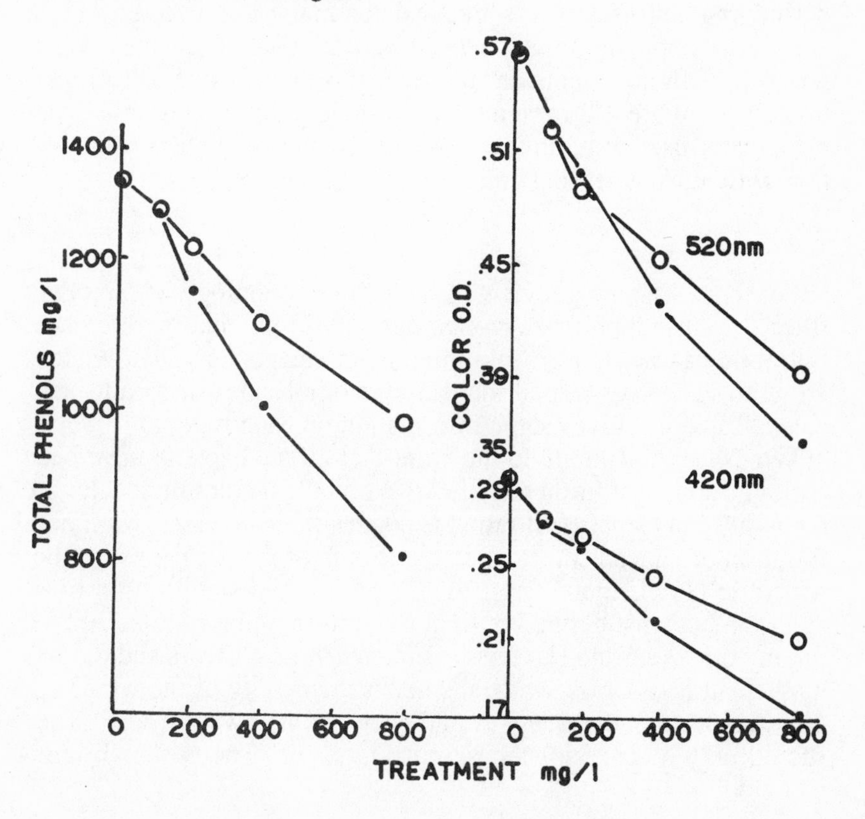

expect in the way of flavor changes. What treatments to give a wine so it tastes the way one wants after two or more years cannot be gained from books or from school. Only through many tastings of fining trials can the fine differences caused be determined and evaluated. Not all finings made may give a positive sensory impact to the wine. Apprenticeship with an experienced winemaker gives an insight into this area of winemaking that is invaluable to the new enologist.

CLARIFICATION AND FILTRATION

Centrifuges

Centrifuges have many uses in a winery. Boulton (1986) summarized the uses for both desludging and decanting centrifuges. The desludging centrifuges are practical for many uses. Among these are clarification of press juices, free-run juice, fermented wine, removal of fining agents or tartrates and yeast (to stop a fermentation). See Figure 4-2 for a modern desludging unit. Decanting centrifuges are used to remove solids from fresh juice before fermentation without first settling and racking (see Chapter 2).

Filter Type

For those wishing a review of work done in enology with modern filters, the handbook by Anonymous (1985) is a handy reference. Filtration can be divided into three main categories. The divisions are removal of suspended matter, macromolecules or small molecules. Table 4-4 gives some information on each type.

Wines can be filtered to the point that all the bacteria have been removed and even further filtered to a point that color molecules are removed. This level of filtration is not desirable in wines. Minimum filtration of wines is most desirable. Some of the occasions to filter wine are: (1) off the fining lees, (2) out of cold stabilization treatment, and (3) just before bottling. Seldom should a wine be filtered for any other reason. The less treatment a wine receives and the less movement it is subjected to, the better the quality of the wine.

Filters are of several types and each has its own purpose. They should only be used for the purpose intended. The two main divi-

FIGURE 4-2. An example of a desludging centrifuge.

Table 4-4. Types of Filtrations and Characteristics.

Process conditions[1]	Filter type		
	Reverse osmosis	Ultrafiltration	Micro or membrane filtration
Pressure used	10 to 60 bars	1 to 10 bars	1 to 5 bars
Particles retained	Ions, sugars organic acids	Polymers, proteins, etc.	Suspended particles
Molecular size, daltons	<500	500 to 500,000	>500,000
Pore size	<1Å	1 to 20Å	22 μm to 65 μm

[1]Filter material can change the characteristics.

sions of filters are "depth" filters and "membrane" filters. The depth filter usually consist of fibrous material compressed into mats or pads and have an electrical charge. These pads are made up of specially prepared cellulose or asbestos. Asbestos is probably the superior filtering material. If asbestos is used, it is important to always filter the wines through a membrane filter before bottling. In fact, no California wineries now use asbestos because of the potential health hazard to the workers in the winery who have to handle the pads.

Depth filters remove yeast, bacteria and other material such as plant fragments, tartrate crystals, etc., from the wine. The tortuous path created by the overlaid fibers traps the particles. Different grades of filter pads are available from very coarse to very fine porosity.

Occasionally filter pads will have undesirable materials adsorbed on the surface. In order to remove these substances, the pads should be washed with a 1% solution of either hydrochloric acid or citric acid. Always put the filter pads in place with the rough side out and the smooth side of the pad against the backup frame. Make enough acid solution equivalent to two times the volume of the pump, filter and lines. Circulate the acid solution through the filter for about five minutes. Wash out the system with 10 volumes of water. The precoat of filter aid (described below) can be put on either before or following the cleaning of the pad. Most commercial pads are already prewashed.

The wine is then filtered with the filter pressure as low as possible. High pressure compresses the filter material and slows the filtration. For best results, filter aid should be added to the wine during filtration. It then builds up on the filter paper or pad slowly and evenly.

The lees left after racking contain a significant amount of wine. Lees filters used to recover this wine are of two general types. There are large plate and frame filters that have an extra volume of space between the plates to accommodate the larger amount of diatomaceous earth required. The other type is a drum vacuum filter that is precoated with filter aid. The drum filter rotates slowly and vacuums the wine up through the filter aid. A knife blade removes some of the filter aid as the drum turns. This exposes fresh filter aid to the lees continually (Figure 4-3). The recovered yeast, tartrates, etc., can be processed. The tartrates are recovered by alkaline extraction. The residue is washed and dried for cattle food. The tartaric acid industry is fairly well controlled internationally by Italian, French and German firms. It is difficult to compete.

The membrane filters are used to sterilize the wine at bottling. The filter is a thin, hard membrane with holes or pathways of a limited size. Any particle larger than the specified size cannot pass through the filter. There are two types on the market. One is a membrane with holes of a uniform size which are made by atomic particle bombardment. The other is a cellulose membrane with controlled pore sizes and a slight depth filter characteristic. The latter is the more popular product. Both require special holders and require that the wine be prefiltered to exclude particles larger than 1μm; otherwise the membrane becomes clogged and is ruined. The membrane pore size normally used is 0.65μm, 0.45μm or 0.22μm. A very tight filtration with the finest asbestos or cellulose pad will also sterilize wine.

FIGURE 4-3. A large vacuum filter used to clean up lees and juice.

Filter Aid

An adjunct to any depth filtration is the use of filter aid, diatomaceous earth. Diatomaceous earth (kieselguhr) is comprised of small sea animals called diatoms. A major deposit is in Lompoc, California, but other deposits are found elsewhere. These consist mostly of silicon oxide with smaller amounts of Al, Fe, P, Ca, Mg, Na, and K oxides.

Postel et al. (1987) looked at 14 kieselguhrs and 7 perlites for metal release when used in wine. The perlites released smaller amounts of Cr, Fe and Al than the kieselguhrs. In all cases, the kieselguhrs raised the Cr, Na and Al and, in some cases, also Cd and Fe. The treatments caused no significant differences in Cu, Zn, Sn, Pb, Ca, Mg, or K.

Filter aids were analyzed by Cirillo (1989) for mineral content. Diatomaceous earth and perlite were compared in Table 4-5. The pH ranges for diatomaceous earth and perlite were 8-8.5 and 7-7.5, respectively. They can be graded into different sizes for slow reten-

Table 4-5. Comparison of Filter Aid Composition.

Component	Diatomaceous earth %	Perlite %
SiO_2	89.00	73.50
Al_2O_3	3.50	21.50
Fe_2O_3	1.45	0.56
CaO	0.50	0.23
MgO	0.07	0.07
$Na_2O + K_2O$	4.50	4.00
TiO_2	0.05	0.04

Cirillo (1989)

tive or more rapid, less retentive filtration. Perlite is a volcanic stone with a similar composition to diatoms. It has slightly less silicon oxide and more aluminum oxide as well as sodium and potassium oxides. The other metals are in trace amounts. It is more abrasive than diatomaceous earth and not as effective.

The purpose of diatomaceous earth is to prolong the use of the expensive filter pads by prefiltration through a "powder" filter (Figure 4-4). This consists of a "paper" pad or fine metal screen which had been coated with filter aid. Alternatively, the regular filter pad can be coated with the filter aid. To be effective, the filter aid should be added to the wine before filtration and kept in suspension during filtration. This enables the filter aid to build up evenly on the filter pad. In this manner, the pad will not be "blinded" or clogged with the material being removed by filtering. When this procedure is followed, a filter pad can be used for large volumes of wine. If too much filter aid is used, the space in the filter will fill up too rapidly. If insufficient filter aid is used, the filter will blind rapidly.

Serrano et al. (1987) compared some American and French diatomaceous earths. The desorbable metal content was very small in all of them. A few did contribute slight off-aromas. There were no great differences among the samples otherwise.

Filtration – Commercial

If wines are to be sterile, the filtration must be effective. Serrano (1988) compared the use of three grades of diatomaceous earth (coarse, medium and fine) to filter a young wine. He also took another wine and prefiltered it through a retentive diatomaceous earth filter. He compared the results of filtration through a medium and sterile pad filter and through a membrane filter. The results are given in Table 4-6 which probably are typical for young wines after finishing yeast and malolactic fermentation.

Plate and frame filters (Figure 4-5) are common to most wineries. Because they can be made to sufficient capacity to filter very large amounts, they are cost efficient. If the capacity is insufficient, the labor cost of breaking the filter down and setting up again becomes

FIGURE 4-4. A powder filter used for coarse filtration of juice or wines.

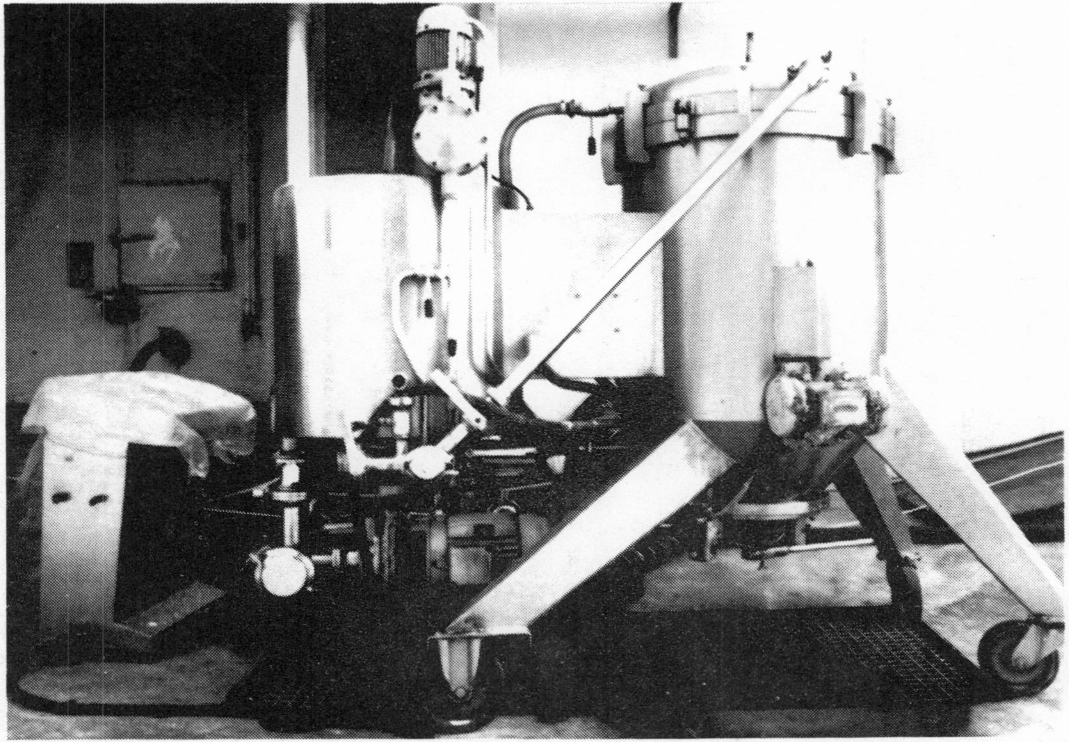

Table 4-6. Filter Tests for Various Grades of Diatomaceous Earth and Types of Filters.

Measurement	Control wine	Diatomaceous earth filters		
		Coarse	Medium	Fine
Flow rate hl/h/m^2	-	20	15	6
Turbidity N.T.U.	60	15	2	0.4
Viable yeast/100 ml	1.8×10^5	5×10^3	2.5×10^3	5.0×10^2
Viable bacteria/100 ml	2.7×10^5	1.2×10^4	3×10^3	1×10^3

Measurement	Prefiltered[1] wine	Filter type		
		Medium pad	Sterile pad	Membrane 0.65 μm
Flow rate L/h/m^2	-	800	400	380
Turbidity N.T.U.	1.6	0.48	0.22	0.24
Viable yeast/100 ml	1×10^2	5×10^1	<1	<1
Viable bacteria/100 ml	5×10^3	4×10^2	<1	<1

N.T.U. = Nephelometer turbidity units.

[1]Prefiltered through diatomaceous earth.

prohibitive. Powder prefilters using diatomaceous earth can also be made to a very large size.

Membrane filters are much smaller and are used only as a final stage after all particles down to 1 μm are removed by prefiltration. The membrane filters then can remove all bacteria and yeast. Sterile pad depth filters can also remove all bacteria and yeast. Salgues et al. (1982) studied the regeneration of membrane filters and the optimization of use. Filter aid, prefilters and fining were all advanta-

FIGURE 4-5. A large plate and frame filter, mainly used for fine filtration of wines prefiltered through a powder filter.

geous to extending filter capacity. They gave some suggested methods of soaking in detergent solutions before back-flushing.

Recently, a new type of membrane filter has come into use. It is named "crossflow" or "tangential" filtration. Ludemann (1987) describes such a system for microfiltration. The unit (Figure 4-6) consists of a series of polypropylene capillary tubes, 2 mm inside diameter and pore size of 0.2 μm. The polypropylene has been rendered hydrophobic. Wine flows perpendicular to the tubes arranged in a suitable container. The wine is recirculated until the desired degree of filtration has taken place. The system can be automatically back-flushed during the run. Wine comes in automatically. A percentage is automatically discharged continuously except for very short intervals when the back-flushing occurs. Effects on wine flavor needs further investigation on these very tight filtration systems.

Ultrafiltration probably is definable as "any filtration which removes materials of a molecular size." Peri et al. (1988) investigated different pore-size membranes (200, 20 and 5 nm) using crossflow filtration. The 200 nm is equivalent to 0.2μm; the 20 equivalent to a 100,000 dalton cut-off filter; and the 5 nm to a 20,000 dalton cut-off filter. The filters consisted of cellulose nitrate, polysulfone and cellulose acetate, respectively. They filtered two wines: one (white table wine) clarified with silica gel and gelatin, and the other (red table wine) settled by gravity only. For the white wine, decreasing pore size removed more colloids; otherwise, little differences were found in the wines. The red wine was not properly clarified by the 0.2μm filter and retained some visible colloids. The 100,000 dalton cut-off filter removed some color and astringency and gave a brilliant wine. The 20,000 dalton cut-off filter removed too much from the red wine. They suggest the optimum pore size between 200 and 20 nm.

Flores et al. (1988) found some pretreatments of white juice increased rates of filtration. The treatment allowed more soluble proteins, pectins and phenolics to pass through the 10,000 dalton cut-off ultra filter. Extended process time increased losses through the membrane of these components. Pretreatment with bentonite and use of SO_2 in the juice were essential for juice clarity.

β-Glucan is a neutral polysaccharide associated with *Botrytis ci-*

FIGURE 4-6. A schematic of "crossflow" or "tangential" type filtration unit. 1. Feed tanks, 2. Feed pump, 3. Heat exchange, 4. Heat exchanges (heating), 5. Circulatory pump, 6. Tangential flow circuit, 7. Capillary cross flow unit, 8. Diaphram plunger for back flushing, 9. Air supply for backflush diaphram plunger, 10. Retentate outlet, and 11. Product tank.

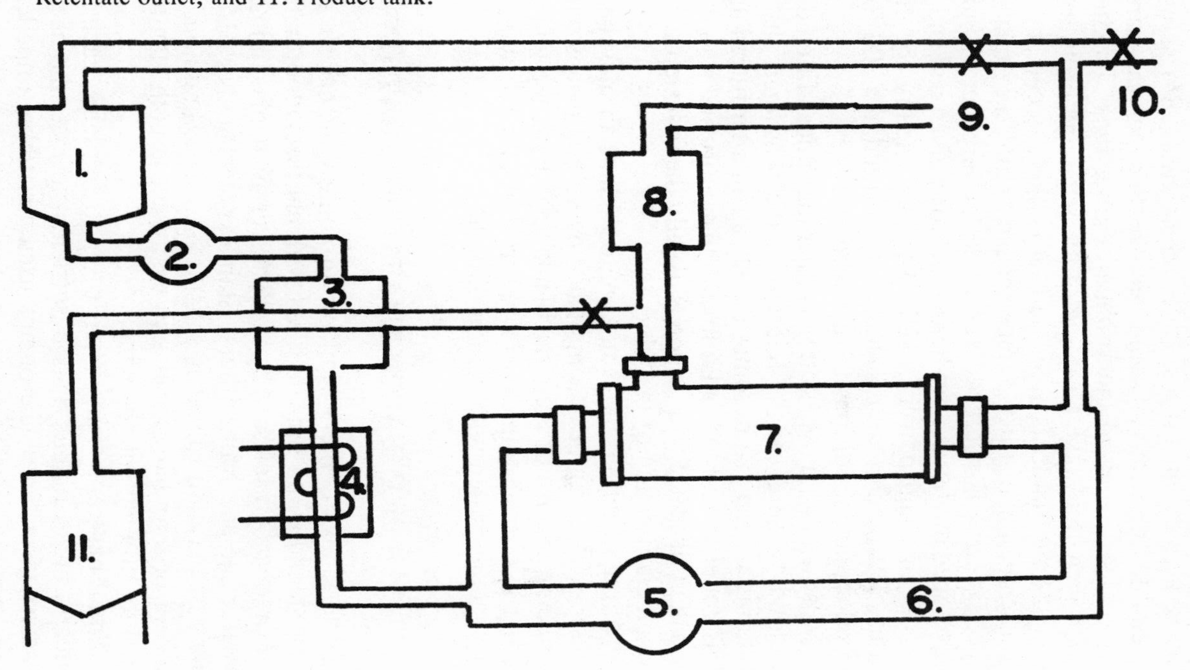

nerea activity on grapes. Villettaz et al. (1984) showed that an available β-glucanase could remove this polysaccharide from a wine. This extended the volume of filter runs. To check for this polysaccharide, take a sample of the wine and bring to 30% ethanol. If long filaments precipitate, it is likely to contain this glucan.

The amount of filter surface needed to filter a certain volume of a specific wine is useful to know. de la Garza and Boulton (1984) developed models for predicting this for a given pad and pump. The exponential and power models were satisfactory for given conditions. The models were successfully verified in a 200-fold scale up to pilot plant conditions from a laboratory test.

Most wines are over-filtered rather than under-filtered. However, a cloudy wine is usually rejected by the consumer. Just repeatedly filtering a wine that becomes cloudy again after filtration is unwise. It is better to find the cause for the instability and stabilize the wine and then filter it. Stabilization is usually the first step after the initial rough clarification of the wine.

During filtration, the same care to exclude air must be taken as in racking. This can be accomplished by several means with the easiest being the use of inert gas blanketing. Rack the wine before filtering if it has a sediment of any sort. Rack a white wine into a tank or container filled with carbon dioxide.

HOME WINEMAKING FILTRATION

The kinds of filters available to home winemakers are limited. One should avoid funnels filled with cheesecloth or cotton as they are practically useless. The best choices available commercially are small plate and frame filters. Stainless steel, either 316 or 304, is the best but hard rubber is satisfactory. Iron, brass and aluminum should be avoided as they can easily cause both stabilization and sensory problems in the wine.

Small plate and frame filters of up to 14 plates, each four inches square, are available. This is sufficient to meet the needs of most home winemakers. The pumps used are rubber impeller type that have positive displacement and can generate sufficient pressure. The number of plates can be adjusted. If only a small amount of wine is to be filtered then fewer plates are used. Now available are cartridge-type filters which can be used effectively on wine. The holders are made of plastic and the cartridges are cellulose. They are made of layers of finer and finer porosity. Cartridges for diatomaceous filtering only are also available. The cost for this cartridge-type of complete filtering equipment, including pump, is reasonable. Sterile filtration can also be achieved with this type of equipment for the home winemaker.

Membrane filter holders are also available for small filtration work for home winemakers. These are relatively expensive compared to the plastic cartridges mentioned above but are available in the appropriate sizes and can be expected to do a superior job. Bottling is covered in more detail in a later section.

There are points that should be stressed concerning the racking and clarification of young wines. Exclude oxygen from white wines as much as possible during handling and storage. Rack the wines on a set schedule to get them off the lees as rapidly as possible. If the wine does not settle clear rapidly, then fine with the appropriate agents used in the proper amounts.

Some deliberate oxidation of red wines is favored. This speeds aging and removes any H_2S. If practiced, it is done after malolactic fermentation. There are many small lots of wine ruined by poor clarification, careless oxidation and later bacterial and yeast infections. It is wise to maintain a free SO_2 level in the white wines of 20-30 mg/l. This will help prevent spoilage and oxidation.

Chapter 5

Stabilization

The precipitation of materials in wine a well as color, aroma and taste changes which occur can either be desirable or undesirable. Wine, until a relatively few years ago, was made very simply. Grapes were crushed. Natural yeast fermented the juice or must. The wine was racked several times and sometimes kept in barrels. The microbiological activity was usually significant and unpredictable. The wine was then bottled. The wine always had sediment, was seldom brilliant and often brown in color. From the 1950s on, all this changed. Wines now must not have sediment, no matter what the storage conditions, and must maintain good sensory quality.

The stability problems that face the winemaker are: (1) oxidation, (2) tartrate, (3) color, (4) hydrogen sulfide, (5) protein, (6) metal, (7) phenols, and (8) microbiological.

OXIDATION

Oxidation changes occur in grapes by the activity of several enzymes. The natural polyphenol oxidase (tyrosinase) is present to some extent in all grapes but the amounts vary. The other oxidation-producing enzyme is laccase but it is present only in mold-infected grapes. Both enzymes will cause oxygen to react with certain phenolic compounds, producing a brown color in the juice. These enzymes are alcohol-unstable and do not survive fermentation. Most of the brown compounds generated in the juice drop out during fermentation. In some juices, this oxidation is encouraged to prevent the color changes from proceeding later in the winemaking process. This has been discussed in some detail in the grape-processing chapter.

The polyphenol oxidase will deplete the oxygen rapidly in freshly crushed grapes.

Under certain conditions, one needs to know how much dissolved oxygen is in a juice. If the SO_2 level is sufficient to inhibit the polyphenol oxidase enzyme, then it is probably saturated. Sadler et al. (1988) determined a standard equation to calculate the solubility of oxygen in grape juice and other sugar solutions at various °Brix and temperatures. Between the range of 4°C and 40°C, the following equation is valid.

$$\ln(\text{mg/L } O_2) = 2.63 - 0.0179°\text{B} - 0.0190°\text{C}$$

where °B = degree Brix, °C = degrees centigrade and ln = natural logarithm.

During the phase of rapid fermentation there is little concern for the exclusion of air as the evolution of the CO_2 effectively does this. However, at the end when the fermentation slows down, air must be excluded especially in small lots. Fermentation traps (described previously) are essential for white wines during this period.

Exclusion of air during racking of young wines has been described in the previous chapter. This practice is good for both red and white table wines and should be followed until the wines are bottled. The oxidation of a sound white table wine or rosé is never desirable. In some cases, red wine may be deliberately oxidized to remove traces of hydrogen sulfide or hasten maturity.

One common problem in commercial white wines is the "pinking" of wines. This is caused by the oxidation of a colorless compound into a colored compound. Usually this appears only in certain varieties; Sauvignon blanc is one in which this occurrence is most noticeable. Before winemakers became fully aware of the desirability of preventing oxidation, this reaction occurred early in the processing and the colored material was removed during early filtrations. Thus, under conditions present during home winemaking, the "pinking" will occur early and will not become a problem after bottling. A prevention for this reaction is the addition of 25 mg/L of ascorbic acid along with sufficient SO_2.

Browning is a problem in some varieties. It is due to the oxidative changes caused indirectly by the presence of molecular oxygen primarily. Occasionally it is caused by other chemical reactions. Oxygen reacts rather slowly with wines. The action of oxygen, chemically, is to form hydrogen peroxide which is highly reactive and can cause the wine to oxidize rapidly. This reaction not only causes browning but also can cause flavor changes. Sulfur dioxide serves the purpose of scavenging out the hydrogen peroxide.

$$H_2O_2 + SO_2 = H_2O + SO_3$$

To be effective as an antioxidant, the SO_2 must be in the free state and maintained at 15-25 mg/L (see Chapter 9 for SO_2 analysis).

Every wine has a natural capacity to tolerate the presence of oxygen. Some wines can absorb large amounts while others can absorb much less. It is best to assume that all wines have a low oxygen tolerance and be a careful winemaker. Red wines are less easily oxidized and have a much higher tolerance for oxygen because of their phenolic content. Rosé wine has less phenolic content and is browned more easily than red wines but less easily than white wines.

PVPP (polyvinylpolypyrrolidone) adsorbs the brown color and phenols when added to white wines. Binding action of PVPP prefers leucoanthocyanins > catechins > flavonols > phenolic acids. The binding is through a hydrogen bond to the oxygen of the ketoimide group of the five-membered ring. This treatment is effective but should only be necessary if the wine has been carelessly made or is very high in phenols, such as in press wines. Moldy grapes made into wine are usually treated with PVPP or casein. Suspend the PVPP in the wine at 6 lb/1000 gal (about 700 mg/L). After 24 hours of settling, filter the wine.

Casein has several advantages for fining white wines. It will combine with some phenols which cause browning as well as the browned complexes. The wines are fresher tasting when fined. The amount used varies from 5.0 to 100 g/hl (50 to 1000 mg/L).

Make a 2% sodium or potassium caseinate in water. To the wine, add from 1/2 lb to 2 lbs/1000 gal (60 to 240 mg/L). The 2% solution is 2 g/100 ml or 20 mg/ml. To get 60 mg/L, add 3 ml of the casein solution to a liter of wine. To the wine (or test sample), add the equivalent amount of tannin equal to half the casein to be added (by wt.). Make up the tannin solution in 10% ethanol v/v as 5% w/v of USP tannic acid. The calculation of the amount of tannic acid solution to add, if you used 3 ml of the 2% casein solution, would be

$$\frac{3 \text{ ml casein sol.} \times 2\% \text{ casein sol.}}{5\% \text{ tannic acid sol.} \times 2} = 0.6 \text{ ml of } 5\% \text{ tannic acid}$$

Set up a series of test samples over the suggested range. Use the one with the least tannin giving the best results after settling. It is seldom necessary to go to higher levels. There is little chance of overfining as the tannic acid is calculated to remove most of the casein. If a considerable excess of casein is used, significant casein will remain in a white wine (Watts et al., 1981). This should not be considered harmful.

INORGANIC AND ORGANIC PRECIPITATES

The materials which cause the most instability and cloudiness in wines are listed in Table 5-1. Also listed are tests to identify the precipitates and a stability test to apply to the wine. Other tests are given in Amerine et al. (1980).

Tartrates

Tartrate deposits in wine are usually crystalline and easily recognized. They do not hurt the quality of the wine and may not be a problem for most home winemakers, but in commercial wines the excess tartrates must be removed.

Grape juice is a super-saturated solution of potassium bitartrate (KHTa). This compound is very insoluble in water and even less soluble in water-alcohol solutions. The KHTa can slowly precipitate over a long time. The rate of precipitation is a function of the temperature, the number of points of crystallization, and the motion of the wine over the crystallization points and the amount of bitartrate present.

The old general procedure was to cool the wine to near its freezing point for 10 days or so. Figure 5-1 shows a typical tube-in-shell

Table 5-1. General Stability Identification and Tests.

Component	Test of precipitate	Stability test[1]
Tartrates	KHTa-Soluble in hot water Tastes acidic CaTa--Insoluble in hot water	Store at -3° for 14 days (KHTa ppts. in 4 days CaTa takes longer to ppt.)
Protein	Insoluble in dilute HCl Redissolves on heating to 80°C	Heat 48 hrs at 120°F. Cool and observe for ppt.
Iron	Soluble in cold dilute HCl. Addition of sodium dithionine immediately dissolves it.	Aerate vigorously and store at 0°C for a week.
Copper	Soluble in cold dilute HCL. Aerate and hold for 24 to 48 hrs, wine becomes clear. React with biquinolyl after dissolving with acid and oxidation.	Exposure to indirect sunlight in a clear bottle for 7 days or heat to 30°C for 3 to 4 weeks. A quick test is to expose wine in a clear bottle to U.V. light for a few hours and observe the cloud--longer tests are more appropriate.

[1]Cloud or ppt. forms

heat exchanger for cooling the wine. This allowed the KHTa to settle out. Along with the KHTa precipitation, cold unstable proteins, color and some calcium tartrate also settled out. Air must be carefully excluded from the cold wine. The solubility of oxygen increases with decreasing temperature. At the low temperature, the oxidation rates are slowed. When the wine is removed from the cold to cellar temperature, oxidation can become a problem.

The wine should be racked or filtered from the cold, not allowed to warm up first so the tartrates can redissolve.

Rhein and Neradt (1979) demonstrated that seeding cold wine with finely ground KHTa crystals allowed for very rapid clarification of wines. The crystals were kept in motion during the process and continually filtered out. Figure 5-2 shows a modern plant for continuous removal of tartrates by this method.

The crystallization rate of potassium bitartrate crystals is dependent on seed crystal size and concentration and the wine movement over the crystal surface (Dunsford and Boulton, 1981a). They (Dunsford and Boulton, 1981b) found effects of cultivars on the rate of crystal growth.

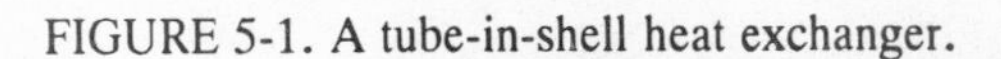
FIGURE 5-1. A tube-in-shell heat exchanger.

Scott et al. (1981) passed the chilled wine through a bed of finely ground KHTa. A rapid tartrate stabilization of the wine was achieved.

Rapid tartrate removal by seeding can be aided by first ultrafiltering the wine. This removes the colloids that can block the nuclei sites and delay or prevent complete stabilization. Conductivity measurements of a wine during the stabilization process can show the effectiveness of the treatment (Bott, 1986) (Figure 5-3). Wucherpfennig et al. (1987) also verified that colloids can fill the crystal sites of potassium acid tartrate and slow precipitation by the "contact" method.

Using conductivity measurements, Wucherpfennig et al. (1988) also found that an excess of calcium ion can foul the tartrate crystals used for seeding by the contact method and slow precipitation of the potassium bitartrate.

FIGURE 5-2. Continuous bitartrate stabilization unit with crystal seeding and agitation (Stellenbosch Farmers Winery, South Africa).

FIGURE 5-3. Schematic of the effect of various stability treatments on the relative conductivity of a wine.

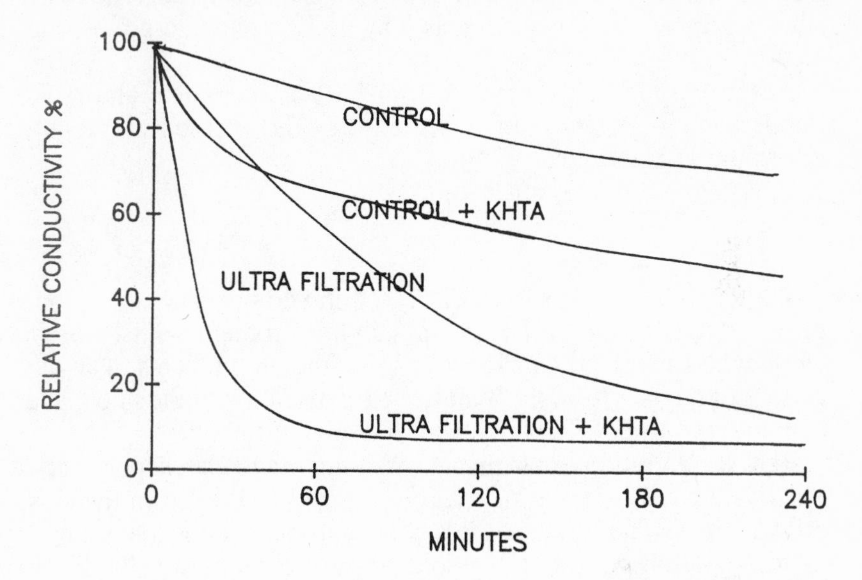

Some colloids can stabilize tartrates for a long time (such as carboxymethyl cellulose). O'Brien (1986) tested the use of carboxymethyl cellulose on the inhibition of KHTa precipitation in some Australian wines. He found it effective as a stabilizing agent. This reacts as other colloids or metatartrate to block the nuclei and prevent crystal growth. Further long-term testing is needed to determine if the wines will remain stable or, as with metatartrate use, destabilize after a few months.

The calcium tartrate is usually not a problem in table wines unless calcium has been added to the wine. This can occur by *not* washing the filter pads or by adding calcium carbonate to lower excess acidity. Usually 50-100 mg/L of calcium is a safe stability level.

Testing by Escudier et al. (1987) showed that crossflow ultrafiltration resulted in wines much easier to KHTa stabilize. Red and white wines had similar kinetic constants. The colloid retention by this system was attributed to the better results allowing 5-8°C contact temperature savings.

The addition of D(-)-tartaric acid or D,L-tartaric acid can cause major stability problems. As little as 10 mg/L of calcium will cause slow precipitation. It may take as long as 12 months to go to completion.

According to Wucherpfennig et al. (1987), complete and rapid stabilization of calcium can be achieved by the addition of the salt, potassium D,L-tartrate.

Color

White wine color has been discussed in various paragraphs concerned with oxidation. The main stability problem with color in white wines is directly attributable to oxidation problems. The exclusion of oxygen from the wines, the preoxidation of the juice, and the judicious use of SO_2 are all preventative means.

Red wine color is due to anthocyanins and related polymers. Most of the red color in *Vitis vinifera* grapes is located in the skin cell vacuoles. The release of this color into the wine is due primarily to the killing action of heat and ethanol on the skin cells. Refer to Chapter 2 for a discussion of the anthocyanin molecules.

V. vinifera varieties only have one glycoside attached to the anthocyanins. *V. labrusca* and other *Vitis* varieties, with rare exception, have some diglucoside pigments. This is one way to differentiate the species.

The addition of SO_2 causes a portion of the anthocyanins to be in a colorless form. Also, the pH of the wine causes changes in the percent of "ionized" anthocyanin compounds, hence, the amount and hue of color present. The lower the pH, the "redder" the wine, and the higher the pH, the "browner" or more "purple" the wine.

All red and some rosé wines will have a color precipitate form during cold stabilization. This will be filtered out. No red wine will remain color-stable for extended periods. Eventually the anthocyanin molecules polymerize with themselves and other phenolic compounds and form insoluble precipitates. Almost any red wine over 5 to 10 years in a bottle will form some "mask" on the bottle. Somers (1988) summarized the knowledge in this area. Several polymer types have been suggested as the stable color or eventual precipitant. None have been characterized yet.

There is no real test for color instability in a wine outside of a cold precipitation test. This test will be positive on most deeply colored wines unless freshly cold stabilized. Low pH and reasonable SO_2 help stabilize, also. Cold precipitation is useful with careful filtering while cold. The closer to the freezing point of wine (−2 or − 3°C), the more polymerized color pigments can be removed by this filtration.

Hydrogen Sulfide

Hydrogen sulfide is a gas formed by the reduction of sulfur.

$$S + 2[H] \longrightarrow H_2S$$

The usual cause is the improper use of dusting sulfur on the grapes to prevent mildew (oidium). This practice should cease in the vineyard at least 6 weeks before harvest.

The odor can be a definite, strong, yeasty odor to a very disagreeable rotten-egg odor. If it is present during fermentation, it will be in the wine. In Chapter 2, methods for control of dusting sulfur in the juice are given. On the other hand, H_2S may come from the absence or low level of amino acids in the juice. If this is the cause, then the addition of about 50 mg/L of ammonia or more is suggested. This is usually accomplished by using DAP [diammonium phosphate-$(NH_4)_2HPO_4$], 1 lb/1000 gal is equal to 32.7 mg/L of ammonia.

The removal of H_2S from wines can be managed by addition of copper sulfate. The excess copper then must be removed by Cufex® (Metafine®). This is a proprietary ferrocyanide treatment which can be potentially dangerous in the hands of amateurs. The procedure is not recommended for non-commercial operations.

Copper treatment of wines to remove H_2S is done by adding 4 g of $CuSO_4$/1000 L of wine. An easier treatment for the amateur is to aerate the H_2S wine to obtain at least partial oxygen saturation (shake up wine vigorously for a minute in contact with air). Add 25 mg/L of SO_2 and let stand (in closed container) for one week to 10 days. Then filter the wine. This usually will remove the H_2S, but not mercaptans or disulfides. The oxygen oxidizes the $S^=$ to sulfur and the SO_2 will reduce the $S^=$ to sulfur.

$$2S^{=} + SO_2 + 4H^{+} \rightarrow 3S^{\circ} + 2H_2O$$

$$S^{=} + 1/2\ O_2 + 2H^{+} \rightarrow S^{\circ} + H_2O$$

The elemental sulfur settles and precipitates and can be removed by a tight filtration. If it is not removed and the wine becomes anaerobic, the $S^{=}$ will be formed again. If yeast are present, they also can reduce SO_2 to $S^{=}$ under the anaerobic conditions in wines.

The best test for the presence of H_2S is the nose. Chemical tests are less than adequate compared to sensory acuity.

If the H_2S is not removed promptly from the wine, it may be converted to ethyl mercaptan. This compound is extremely difficult to remove and its removal generally decreases the wine quality to a great extent. This compound, C_2H_5SH, has a cabbage-like, skunk-like or treated natural gas smell and is very disagreeable. Oxidation of ethyl mercaptan to diethyl disulfide can occur.

$$2C_2H_5SH + 1/2\ O_2 \rightarrow C_2H_5\text{-}S\text{-}S\text{-}C_2H_5 + H_2O$$

Goniak and Noble (1987) determined the threshold of the various volatile sulfur compounds that might be found in wines. These were dimethyl sulfide, dimethyl disulfide, diethyl sulfide, diethyl disulfide and ethanethiol (ethyl sulfide). The corresponding threshold values were 25, 29, 0.92, 4.3 and 1.1 μg/L. The methyl sulfide, not reported in the above study, has been found in wines treated with certain pesticides. This compound has an even lower threshold than the ethyl sulfide.

Protein

There are half a dozen or so protein compounds which are produced by the grape and are found in the fermented wine. These are among the compounds which cause haze in wines along with some glycoproteins and polysaccharides. The short peptides produced by the yeast are not involved in this problem. Some of these protein materials of the grape are precipitated by cold and some by heat. The cold-unstable ones are usually filtered out during cold stabilization and are seldom a problem. The heat-unstable proteins will become unstable in time, if not removed, and cause a haze in most wines. The precipitates of the proteins are non-crystalline. They are usually brown in color and found only in white wines. Red wines, because of their high phenolic or tannin content, cause early natural precipitation and removal. However, they are present to some extent in the juice of red grapes.

There are several ways to remove natural protein from a wine. One is to flash pasteurize the wine to 160°F (71°C), then chill and filter it. This is still occasionally used on dessert wines, especially muscat types which require excess bentonite treatment. This was a common method a few years ago for all wines, but the more gentle treatment of bentonite fining has replaced the pasteurization treatment for table wines. Bentonite fining was described in Chapter 4. One fining is usually sufficient to make the wine protein stable. However, this cannot be taken for granted and tests to confirm it should be made. Silicon dioxide (Kieselsol), especially prepared in finely divided particles, is also effective in removing certain proteins. It can be used more effectively than tannin in gelatin coprecipitate finings. Bentonite is the more effective and preferred treatment for protein removal.

There is a simple test for protein stability. A filtered-brilliant sample of the wine is placed at 120°F (49°C) for 48 hours, then brought to room temperature. At the end of that period, the sample is checked for precipitates and clarity. If the sample is brilliant and no precipitate present, it is judged to be protein-stable. If it is hazy or has a precipitate that does not disappear at room temperature, it is protein-unstable.

Dubourdieu et al. (1988) compared six different protein stability tests. The tests were: 5 min heating at 80°C in a double boiler, 10 days at 35°C in oven, addition of 5 g/L tannin, addition of phosphomolybdic acid (Bentotest®), addition of trichloroacetic acid and chromatographic separation. They concluded that 30 min at 80°C was more efficient but the cloudiness determined after 10 days at 35°C was important. The trichloroacetic acid and Bentotest® were acting on all the protein, not just the heat unstable ones. Tannin addition caused turbidity and did not correspond well to the chromatographic tests.

Hsu and Heatherbell (1987) looked at the selectivity of bentonite for the removal of proteins. The 12,000 and 20,000-30,000 dalton ones were first removed. The higher molecular weight proteins,

60,000-65,000 dalton, were most difficult to remove. The heat-sensitive proteins were less than 30,000 dalton. The glycoproteins were most important. Further work was done by Hsu et al. (1987). Ultrafiltration, using 10,000-30,000 dalton cut-off filters, held back 3-20 mg/L of protein. The 10,000 dalton cut-off filter retained 99% of the total protein under 30,000 daltons. If the wine was still heat unstable, it took 80-95% less bentonite to stabilize it than it took before filtration. They concluded the heat test for stability used in Australia (80°C, 6 hrs minus 4°C 12 hrs) or the California test (120°F (49°C), 48 hrs cool) were superior to the Bentotest® used in Europe.

Miller et al. (1985) measured the effect of bentonite fining and ultrafiltration on wine aromas. They found adding bentonite to the fermenting juice removed proteins effectively. It was more efficient than ultrafiltration or heating the wines. The differences in the aroma constituents were relatively small. The experiment was on a commercial juice and was non-replicated.

Metals

There are three common metals (copper, iron, aluminum) which can cause problems with wines. The first two are still a problem commercially but not as much as when many bronze valves and iron fixtures were used in wineries. Aluminum should never be used for storage of wines. Other metals are just not present in sufficient amounts under normal conditions to be a problem. Be aware that grape juice and wine contain a medium strong acid (tartaric) that can slowly dissolve most metals. This can sometimes cause enough metal pickup for the wine to become undrinkable. Canned wine is seldom seen for this reason. Despite the improved technology, the corrosiveness of wine still prohibits successful metal packaging. Galvanized buckets should be avoided or any other metal container except stainless steel which resists the wine's action.

Metal in crushers and presses can be of iron and valves of bronze as long as they are washed carefully to remove any juice after each use. No juice should remain in contact with the metal for extended times (overnight). If juice does remain in contact with the copper or iron, it will dissolve some. This concentrated metal solution will be picked up in the next day's juice. Even this is not serious if not too much metal is picked up.

Thoukis and Amerine (1956) showed that appreciable copper and iron were adsorbed onto the yeast from juice. Langhans and Schlotter (1987) quantified the amounts of copper that could be removed from a wine by adding fresh or dried yeast. The level of copper could be reduced by 50-75% by the treatments. The dry yeast was more effective than the fresh. This treatment could be substituted for blue fining. The optimum contact time was from 1 hr to 1 day. After the maximum removal time, the copper begins to resolubilize.

The real problem is metal pick-up beyond the fermentation. This can occur from tools falling into the wine or containers or equipment containing these metals.

Aluminum causes the SO_2 in the wine to be reduced to H_2S as well as dissolving in the wine. There is no test for aluminum practical for the home winemaker except his nose. It only takes a relatively short exposure for the wine to develop H_2S and wine stored in this manner is usually beyond rehabilitation.

Iron reacts differently in white wine than in red wine. In red wines, the turbidity caused is called "blue casse." This is a ferric tannate colloid which gives the red wine a metallic blue color. White wines form ferric phosphate or "white casse." The ferric phosphate only forms in a certain pH range (2.9-3.6). Addition of a small amount of strong acid such as hydrochloric acid will cause the grayish deposit or turbidity to redissolve. This is a test for the presence of white casse.

To test for future iron turbidity problems, add a few drops of hydrogen peroxide to the wine, aerate and check the next morning for turbidity. (The peroxide changes the ferrous iron to ferric which is necessary for the precipitant to form.) If no precipitate occurs, the wine is probably iron-stable.

Citric acid is a good protective agent as it chelates iron and prevents colloid formation.

To determine if iron can be successfully stabilized, add 0.7 g/L of citric acid to a wine sample, aerate and add a few drops of hydrogen peroxide. If the wine clouds, a treatment for iron removal should be attempted. If the wine remains clear, it is stable for iron.

The easiest treatment is a casein fining. However, the wine must first be aerated which can cause loss of aroma and flavor. Only white wine is usually casein-fined. (See previous section on casein fining.)

The use of calcium phytate or blue fining (potassium ferrocyanide) is recommended only for the trained enologist. It is difficult to use accurately and can be dangerous.

An amateur may want to fine metals using the compounded ferrocyanide treatment. He should first contact as person experienced in such work and learn the required techniques and dangers involved (see Ough and Amerine, 1988).

The commercial winemaker should analyze the wine to establish the iron level. The safe level is less than 10 mg/L, and most prefer a 5 mg/L level. Most commercial wines average between 5 and 10 mg/L. If the level is safe, no further concern need be had about the wine in that regard.

Fining with Metafine® or Cufex® are accepted treatments. This is a modified blue fining method. The supplier's instructions should be closely followed.

The Hubach test (Ought and Amerine, 1988) will show if a wine has been overfined. If the test is positive, set up a series of test samples of the wine. Add varying amounts of copper. Determine the amount that just leaves a trace of copper in the wine. Add this amount to the tank of overfined wine. Test again for cyanide and for copper to determine if the wine is both chemically stable and medically safe. Copper can cause very disagreeable hazes in the wines. The reaction can be with either protein or tannins or both. Copper mainly comes from grape sprays, addition of $CuSO_4$ for H_2S removal or bronze valves. A quick test for copper can be made by adding a few drops of sodium sulfide solution to 20 ml of wine and comparing that tube to one with no sodium sulfide added. If more than 0.5 mg/L of copper is present, a turbidity will form in the sodium sulfide-treated tube.

If the wine has a copper problem, the winemaker has several choices. He can treat the wine with one of the compounded blue fining agents. He can treat with yeast cells to adsorb the copper.

Recently Lobser and Sanderson (1986) have developed a chelating resin that effectively removes both iron and copper from wines. The resin can be regenerated. The chelating substance used was 8-

hydroxyquinoline and can be used in a column. The effective rates at 20°C were between 5 and 10 bed volumes/hr to remove from 70 to 98%. At pH 3.5, the effective capacity was 1.2 mmol Cu^{+2}/g dry resin, 0.5 mmol Fe^{+2}/g dry resin and 1.0 mmol Fe^{+3}/g dry resin.

Phenolic Precipitations

Phenolic material in white wines occasionally will cause some problems when other constituents and conditions are correct. Removal of excess phenols can be accomplished by treatment of white wine with PVPP followed by a bentonite treatment and a low dose of citric acid. This kind of problem usually occurs when bottling a white wine which has passed all the normal stability tests. Occasionally a flavonol can cause problems. Somers and Ziemelis (1985) identified quercetin as the cause of a yellow haze and yellow precipitate in a white wine. The cause was determined to be due to excess leaves in the juice because of poor mechanical harvesting. Leaves contain relatively high levels of quercetin glycoside which hydrolysed during fermentation. Kaempferol was detected but in lesser amounts.

MICROBIOLOGICAL

The proper inoculation of the wine pretreated with 50 to 75 mg/L of SO_2 is sufficient under most conditions to ensure a clean fermentation. Maintaining the wine under anaerobic conditions with adequate free SO_2 (20-30 mg/L) is sufficient to protect the dry wines from most yeast or bacteria. If the wine is exposed to air, then mycoderma, flor yeast or acetobacter will grow. If the SO_2 is not maintained, other bacteria and yeast will grow and spoil the wine, even if air is absent.

Most home winemakers do not have the knowledge or equipment to identify these microbes. However, if anything changes in the wine's appearance either in the barrel or other storage, it usually can be attributed to microbial activity. If a microscope is available, the wine can be observed. Some of the problem microbes are listed in Table 5-2. There are other spoilage microbes but these are the most likely to cause problems.

Table 5-2. Some Bacteria and Yeast Associated with Microbiological Instability in Table Wines.

General term	Genus	Appearance	Cure
Mycoderma	Candida sp.	Surface film--wine	Exclude air
Acetic film	Acetobacter sp.	Surface film--wine	Increase SO_2 Exclude air
Mannitol fermentation	Bacillus sp.	Bacteria in stuck wine--cloudy	Cool--add SO_2 + restart
Tartaric fermentation	Bacillus sp.	Loss of color and acidity--increase in volatile acidity.	Add SO_2 and filter
Butyric fermentation	Bacillus sp.	Loss of glycerin.	Add SO_2 and filter
Flor film	Torulaspora sp.	Crinkly film, increase in acetaldehyde	Exclude air
Malolactic	Lactobacillus sp.[1]	Cloudy sometimes	Add SO_2 and filter
	Leuconostoc sp.	CO_2 evolution	
	Pedioccoccus sp.	Decrease in malic acid	
Ropiness	Lactobacullus sp.	Wine develops a viscous appearance and when poured, resembles egg white in flow character, caused by dextran formation.	Generally beyond cure.
Bretanomyces	Brettanomyces sp.	A definite off taste with the appearance of small yeasts in the wine.	SO_2 and filtration.

[1]A special case is Lactobacillus trichodes, which causes "cottony mold" in dessert wines with less than 75 mg/L of total SO_2.

Heresztyn (1986) isolated *Brettanomyces intermedius, B. lambicus, Lactobacillus brevis* and *L. cellobiosus* from spoiled wine. The first and third were the most vigorous. They were propagated in the presence of 5% ethanol and 5% n-propanol in grape juice. Four compounds associated with the odor of *Brettanomyces* spoiled wines were found: (1) 2-acetyl-1,4,5,6-tetrahydropyridine [I], (2) 2-propionyl-3,4,5,6-tetrahydropyridine[II] and their isomers, (3) 2-acetyl-3,4,5,6-tetrahydropyridine, and (4) 2-propionyl-1,4,5,6-tetrahydropyridine.

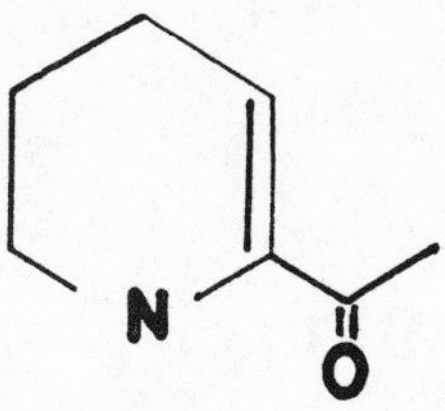

I 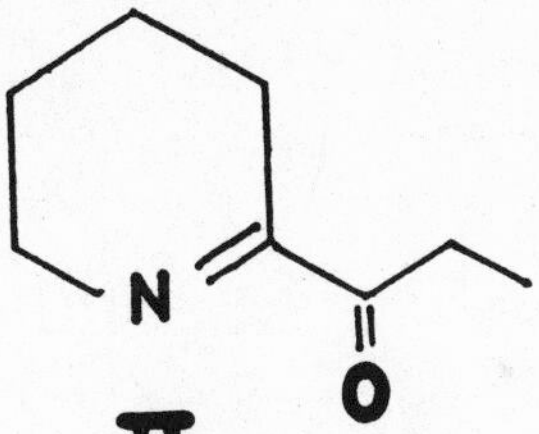 II

Schizosaccharomyces pombe has been used in certain instances as a fermenting yeast because of its ability to remove malic acid. Unterholzner et al. (1988) termed it a spoilage yeast because of the off-flavors it produced. Auriol et al. (1987) studying *Schiz.* found that not all strains degraded malic acid completely. He also noted that the degradation was not linked to growth or sugar as previously thought.

There are instances where secondary fermentations are beneficial. These will be discussed in the next chapter.

Microbiological stability relies on a few simple rules:

1. Add sufficient yeast culture and SO_2 to have a good strong primary fermentation.
2. When the fermentation is complete, remove the wine from the yeast rapidly (except in specified occasions).
3. Store in a cool location.
4. Exclude air.
5. Maintain a reasonable SO_2.
6. Do not add or have residual sugar unless the proper precautions are taken.
7. Always keep everything clean.

If these simple rules are followed, the wine will remain stable (except possibly for a malolactic fermentation).

Chapter 6

Secondary Fermentations

When fermented to dryness, most white wines are relatively microbiologically inactive. Red wines, however, will usually undergo malolactic fermentation without too much encouragement. For most yeast to grow in a wine, it must not contain excessive SO_2 and must contain fermentable sugar. There are exceptions of course.

MALOLACTIC

Until a few years ago, most malolactic fermentations were nature's choice. Some were "good" and some were "not good." Twenty years ago, roughly 75% of commercial red wines and 40% of white wines had undergone malolactic fermentation (Ough, 1971). In a malolactic fermentation, the main activity of interest is the degradation of malic acid in the wine to lactic acid. This bacterial activity decreases the total acidity and increases the pH. The bacteria grow more readily at the higher pH and are fairly sensitive to SO_2. In addition, their growth is inhibited by cold.

Once a wine has undergone malolactic fermentation, time is required for the wine to recover. Grapes from a warm area having less malic acid and a high pH benefit much less, if at all, from this secondary fermentation than grapes from cooler areas with a lower pH and higher acidity.

Strains

The isolation and selection of malolactic strains have taken place over many years. It was known that some strains fermented sugars while others basically grew in dry wine and metabolized the malic acid to lactic acid. The three main genera are *Lactobacillus,*

Leuconostoc and *Pediococcus*. The homofermentative species metabolize some sugar to lactic acid and produce minimal amounts of acetic acid and carbon dioxide. Heterofermentative species, on the other hand, ferment sugar and can form significant amounts of acetic acid, carbon dioxide, some glycerol, and alcohol. They can reduce fructose to mannitol. *Pediococcus* are all homofermentative, *Leuconostoc* are all heterofermentative, and *Lactobacillus* can be one or the other.

The principal malolactic bacteria found in wines are shown in Table 6-1.

Much of the work done in France, Germany and the United States has led to isolated strains that can be cultured and dried for use at the winemaker's convenience. *Leuconostoc oenos* var. is the choice as the most effective species. In the areas of medium-warm climate, the degradation of malic acid per se is of no sensory value. In most cases, the wines naturally have a medium acidity. The use of malolactic fermentation may require some readjustment of the acidity of the wine in some cases. There are three good reasons to use this technique. It gives the wine a complexity derived from the biochemical changes of the malolactic fermentation. It assures that the fermentation will not occur after bottling. It is occasionally used to reduce the acidity of a high acid wine.

Davis et al. (1988) found the strains of *Leuconostoc oenos* were more tolerant of low pH than *Pediococcus parvulus* or *Lactobacillus* sp. The latter was more tolerant to SO_2 and ethanol. None of the

Table 6-1. The Principal Malolactic Bacteria Found in Wines.

Family	Genus	Species	Fermentation of glucose	Shape of cells
Lactobacilliacae	Lactobacillus	plantarum	homofermentative	rods
		casei	homofermentative	rods
		hilgardii	heterofermentative	rods
		brevis	heterofermentative	cocci in wine
Streptococcaceae	Pediococcus	cerevisiae	homofermentative	cocci
		pentosaceus	homofermentative	in pairs or tetrades
	Leuconostoc	oenos	heterofermentative	cocci in short chains

strains produced extracellular proteases. The majority produced esterases. The presence of esterases in a wine could speed the loss of the fruity esters. This may be cause for decline in wine quality after malolactic fermentation in some instances.

Microbiology

The formation of 2-butanol is well recognized as a spoilage product of bacterial contamination of wine. Radler and Zorg (1986) have characterized the enzymes causing this reaction from a strain of *Lactobacillus brevis*. The dialdehydrogenase in the bacteria hydrogenates meso-2,3-butanediol to ethyl methyl ketone. The ethyl methyl ketone acts as a hydrogen acceptor during the glucose fermentation by the bacteria and 2-butanol results. The meso-2,3-butanediol is necessary for the formation of the 2-butanol.

$$
\begin{array}{ccccc}
CH_3 & \text{dialdehy-} & CH_3 & & CH_3 \\
| & \text{drogenase} & | & [H] & | \\
HC\text{-}OH & \longrightarrow & HC\text{-}H & \longrightarrow & CH_2 \\
| & & | & & | \\
HC\text{-}OH & & C{=}0 & & HC\text{-}OH \\
| & & | & & | \\
CH_3 & & CH_3 & & CH_3 \\
\text{2,3-butanediol} & & \text{ethylmethyl ketone} & & \text{2-butanol}
\end{array}
$$

Heresztyn (1986) reported that *Lactobacillus brevis* and *L. cellobiosus* both made acetyl- and propionyl-tetrahydropyridine. These were the cause of "mousey" aromas in spoiled wines.

Kuensch et al. (1974) found that *Leuconostoc oenos* caused arginine to be metabolized to ornithine and urea. The urea was broken down to ammonia. Very large amounts of ornithine were formed, from 23 mg/L to 632 mg/L in wine. On a molar basis, this was equal to the amount of arginine metabolized and the amount of ammonia formed. This activity occurred after the malolactic conversion was complete. These changes cannot be related to cellular breakdown and release of components because the arginine was exocellular.

Weiller and Radler (1976) looked at many malolactic bacteria found in wine. They determined the use of the various amino acids by the bacteria. All the *Leuconostoc oenos* and *Lactobacillus brevis* strains tested used arginine up completely except one strain of *L. oenos* which used only 20%. It was the main source of nitrogen for *L. oenos*. *Pediococcus cerevisiae* did not use it but preferred either histidine or tyrosine. The *Lactobacillus brevis* were not fastidious and used many amino acids. One strain of *Lactobacillus buchneri* was similar to the seven strains of *L. brevis*. All those that metabolized arginine gave ammonia and ornithine.

The formation of histamine during malolactic fermentation has been questioned in recent years. Weiller and Radler (1976) could only find one strain out of 105 that could produce histamine.

Surveys of California wines showed no correlation between histamine and the malolactic fermentation in the wine (Ough, 1971). Trials done over many years on experimental wines using *Leuconostocoenos* and *Lactobacillus* cultures did not produce any histamine even in wines fortified with histidine. A recent work of Ough et al. (1987) also did not demonstrate any production of histamine in model wine solutions from several bacteria. Some of the earlier reports that showed correlations with malolactic activity in the wines may have measured some other compound.

Cabanis (1985) reviewed the toxicity and physiological effect of histamine. He concluded that the amounts found in wine were not dangerous. He discussed a report that said an adult could ingest 165-200 mg of histamine and not show the initial response of a flush. Toxicity intravenous or through the lungs is, however, quite dangerous. Histamine causes expansion of the surface blood vessels, lowers blood pressure and ultimately causes a person to go into shock. Tyramine, on the other hand, is a vasosuppressor and causes higher blood pressure, headaches and other vascular problems. The amounts of these compounds normally found in wines are well below any possible reaction-causing levels. Cheeses, for example, have much higher levels of these compounds. When taken orally, these are metabolized rather rapidly by the body unless a person is on prescribed monoamine oxidase inhibitors. These amines are believed to be formed in wines in minor amounts from the corresponding amino acids by decarboxylases.

Inhibition

Malolactic bacteria are inhibited by SO_2. Lafon-Lafourcade and Peynaud (1974) confirmed earlier work and winemakers' observations that bound SO_2 also was effective in preventing growth. They found that free SO_2 was about 5-6 times more effective than the bound SO_2. Pyruvate-SO_2 was slightly more effective than acetaldehyde-SO_2. They found at pH 3.3-3.5 in artificial medium, bound SO_2 levels of 45 and 104 allowed only slight growth and no fermentation of *Leuconostoc oenos*. Acetaldehyde per se was not inhibitory. The bacteria metabolized the bound aldehyde and pyruvate and released the SO_2 in the free form. This apparently was related to the inhibitive effects.

Fumaric acid was used commercially for some time to inhibit malolactic fermentation. Relatively small amounts would effectively inhibit growth. However, the rate of solubility of the fumaric acid was very slow. The effective wetting agents were not approved for use in foods. It was used as an acidulant in place of citric acid at bottling. Wagner et al. (1971) found that it was metabolized during fermentation by yeast in grape juice. They suggested the fumaric acid was converted to malic acid by fumarase. While ATP inhibits this reaction, the inhibition is overcome if fumaric acid is present in saturation amounts or more.

Radler (1986) reviewed the literature on malolactic microbiology. Twenty-two fungicides were investigated for their inhibitory effects on several strains of *Leuconostoc oenos*. Haage et al. (1988) found only Euparen® and Aktuan® to have any significant effects. Inhibition levels were around 1 mg/L and the lethal levels were 128 and 60 mg/L, respectively. With other lactic acid strains as low as 0.125 ppm was inhibiting by Aktuan® and Euparen®.

Growth and Viability

The wine to be inoculated with bacteria should have a pH between 3.25 and 3.50. The total SO_2 should be less than 30 mg/L with essentially zero free SO_2. The bacteria cell concentration in the wine should reach at least 10^6 cells/ml to 10^8 cells/ml.

If the bacteria does not grow, there could be several good reasons: (1) the added culture or dried cells were no longer viable; (2) the

SO_2, pH, ethanol or nutritive condition were such that growth could not proceed; or (3) the inhibition caused by natural buildup of inhibitors or something added extraneously.

Bacteriophages have been found in wines undergoing malolactic fermentation. These phages are specific for a bacteria. These organisms consist of several proteins and either a single RNA or DNA molecule. One of the proteins is a lysozyme which allows the phage to penetrate the host cell. Once inside, the DNA or RNA chain is released and the viral chromosome is replicated. Viruses synthesize new coats and enzymes, break out of the cell and are ready to attack other cells. Multiplication time is short, less than one hr at optimum temperature.

Nel et al. (1987) characterized a number of bacteriophages isolated from wine and sugar cane that infected some strains of *Leuconostoc oenos*. In particular, the commercial strains ML34 and PSU-1 were lysed by many different phage strains. They did not find these phages in grape musts, fresh grapes, ground grape skins, vine leaves or plant treatment water.

Some yeast stimulate malolactic bacteria activity and others inhibit it (Lemaresquier, 1987). The exact nature of these activities have been speculated on but not resolved. Krieger et al. (1986), using four different *L. oenos* strains, found with white wine that alcoholic fermentation inhibited bacterial growth and activity. In a sample of similar juice treated with Pimaricin® to inhibit the yeast growth, the bacteria grew and fermented well. They recommended addition of bacteria after alcoholic fermentation. On the other hand, Beelman and Kunkee (1987) show simultaneous additions of yeast and malolactic bacteria were successful. Providing sufficient malolactic bacteria were present, both fermentation finished with no problems. The sensory characteristics of the wines were satisfactory.

Joyeux and Lonvaud-Funel (1985) investigated several commercial malolactic freeze-dried preparations. They found all the *Leuconostoc oenos* preparations were very satisfactory. It was thought essential to reconstitute the bacterial preparation in grape juice. All fermentation followed were essentially done in 21 days, including those in white wines.

Costello (1987) discussed the use of the proper malolactic bacteria to use for white wine. He suggested selection of a bacteria that had been isolated from a white wine that had undergone malolactic fermentation. He indicated bacteria that do well in red wine may die when used for white wine. The selection should be *L. oenos*.

When propagating commercial starters, the essential reactivation stage of culturing the bacteria cannot be omitted (Costello, 1988). This is usually done at pH 4.5-5.0 at 20-25°C for 48 hrs. Manufacturer's directions should be followed explicitly. The malolactic bacteria is very sensitive to SO_2 so care should be taken to maintain as low a level as possible before inoculation and during the bacterial activity. Inoculation levels should be about 10^5-10^6 cells/ml.

The viability of cultures is not too difficult to determine by several bacteriological methods. The lack of viability of a freshly grown wet culture, or for that matter dried bacteria supplied by reputable suppliers, is seldom a problem. The wine composition should be within the range given. Any significant variation in juice composition from suppliers' limits could be a problem. The upper pH range does not inhibit but higher pH values cause some undesirable microbiological activity and decreased quality.

Lonvaud-Funel et al. (1988) found that fatty acids inhibited growth of malolactic bacteria. As with yeast, addition of yeast ghost cells decreased inhibition.

If one wishes to try a natural malolactic fermentation, the best approach is to use about 50 mg/L or less of SO_2 before the initial yeast fermentation and refrain from further SO_2 addition. At the finish of the yeast fermentation, or after the first racking, place the wine in full containers in a room at 20-25°C. If the malolactic fermentation does not happen within a month, rack the wine and proceed with normal cellar care. Sometimes they will ferment later. It is a gamble to depend on nature for a malolactic fermentation.

Some wineries have a natural malolactic flora established that causes successful malolactic fermentations, but the natural flora in a winery may be a very poor strain that causes off-odors and tastes.

Cultures or slants of the bacteria can also be obtained from some enology departments. These must be cultured specially with sterile tomato juice or grape juice with pH adjusted to about 4.5. After the culture is growing well, an inoculum of the bacteria can be added to a small portion of the wine to be fermented. When active growth appears, that whole starter is added to the bulk of the wine. (Only about 2% of the volume of the wine should be starter culture.) Then hold the wine at 20-25°C until the fermentation is finished.

It is necessary to determine when the malolactic fermentation has taken place. Sometimes it will occur along with the yeast fermentation, other times only after inoculation with bacteria, and sometimes it will not go at all. Below is a satisfactory method for determining if the malolactic fermentation is complete.

Obtain a wide-mouth gallon jar or an appropriately sized chromatographic jar and a package of #1 Whatman filter paper in large sheets. Cut a sheet into 8″ × 10″ size. Mark the paper (with a pencil – inks will run as the solvent gets to that area) and put a small drop of the wine on the paper as shown in Figure 6-1. Dip a capillary tube into the wine to be tested and barely touch the capillary to the intended spot. Allow the spotted drop to dry before the next application. Keep the drop application area as small as possible (1/4 inch in diameter). Repeat the application twice. Apply a control spot of known malic acid using a 3 g/L solution in an adjoining marked area. Transfer 50-70 ml of a chromatographic solution. This consists of n-butanol-formic acid solution saturated with water to which bromcresol green has been added (see Ough and Amerine, 1988). Roll the paper into a cylinder and staple edges together. Place the roll in the jar with the end of the roll closest to the dried applications into the solution. Close the top of the jar with a tight covering. After 4-6 hours, or when the solution has migrated nearly to the top of the paper, remove the paper from the jar. Unroll and hang the paper up to dry. When dry, the control spot of malic acid will be yellow against a blue-green background. If the test spot at the same Rf as the standard is yellow, there is still malic acid present. The completion of the malolactic fermentation is indicated by the absence of malic acid in the test spot. There will be an increase in the size of the top spot (lactic-succinic) on the paper.

Just an increase in the top spot (lactic-succinic acid) is a positive indication of malolactic bacterial activity. The finish of the malolactic fermentation is not confirmed until the test malic acid spot disappears. This test should be repeated on a weekly basis until the fermentation is complete.

When the fermentation is complete, the free SO_2 should be ad-

FIGURE 6-1. Schematic of an organic acid paper chromatogram showing base line, application points, spots and solvent front. 1. Standards, 2. Partially metabolized malic acid, 3. Fully metabolized malic acid, and 4. Malic acid not metabolized.

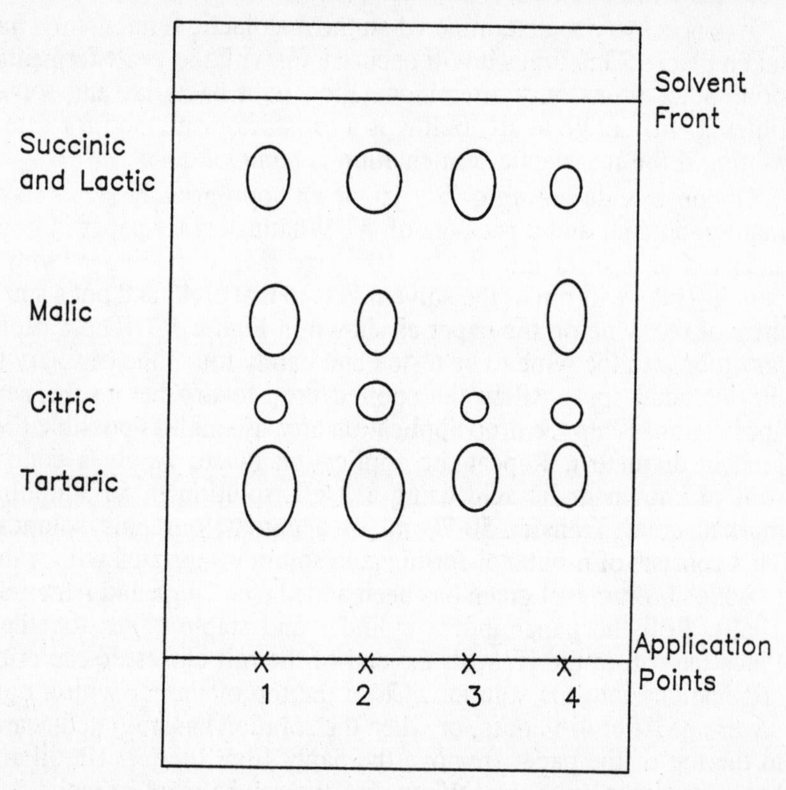

justed to 20-30 mg/L. The wine is then fined and filtered. If cold stability is desired, the wine should be chilled to less than 0°C (32°F). When the wine has been filtered after the desired length of cold treatment, it is then ready for aging.

FLOR SHERRY

To make "flor" sherry by the film method is easy but it requires patience. It may take several years or more to produce a properly aged and flavored product. Flor sherry is made by using a yeast which grows as a film on the surface of the wine, allowing the yeast to make biochemical changes in the wine. Figure 6-2 shows a film growing on wine in a glass-ended barrel. The most obvious change that the flor yeast make in the wine is the production of acetaldehyde. This is produced when the yeast enzymatically oxidize alcohol in the wine. Other flavor and aroma changes also take place. By careful aging and barrel maintenance, a good production can be obtained by the home winemaker.

FIGURE 6-2. A barrel with a glass end showing the flor yeast film on the surface.

Blending and Fortification

The following formula is used for blending or fortification. This formula is set up as a proportion.

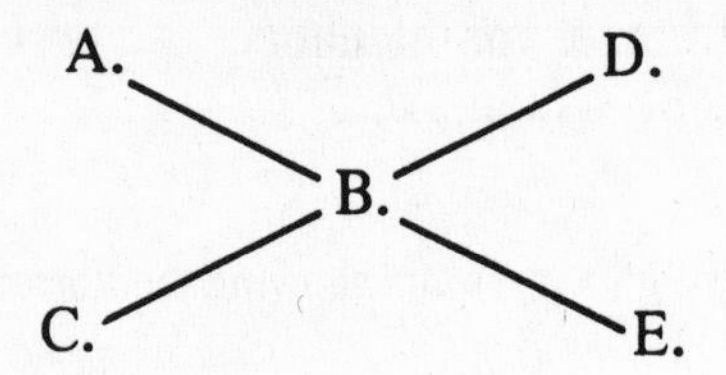

A = Sherry material to be fortified % v/v
B = Desired ethanol % v/v
C = Ethanol fortifying material % v/v
(Proof = 2X alcohol are as % v/v)

Example:

If the wine is 13% v/v = A
Desired ethanol level 15.5% v/v = B
Using high proof at 190° proof or 95% v/v = C

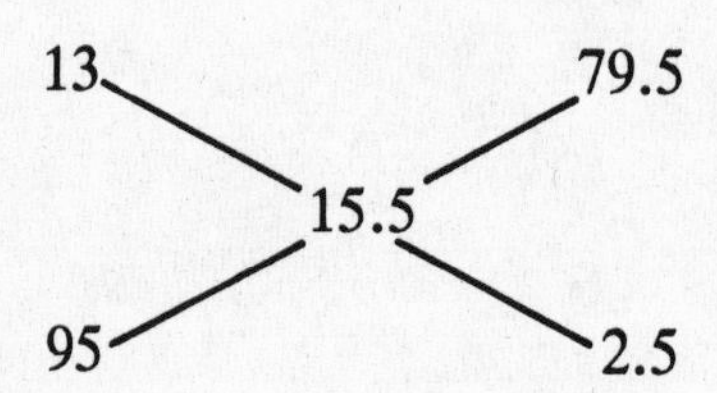

E = B − A = 2.5% v/v
D = C − B = 79.5% v/v E + D = 82.0% v/v

If you want 20,000 gals of 15.5% wine.

$$\text{Wine (unfortified)} = \frac{\text{(Desired gals total) D}}{\text{E + D}} = \frac{(2 \times 10^4)(79.5)}{82.0} = 19{,}390 \text{ gal. wine}$$

$$\text{High proof needed} = \frac{\text{(Desired gals total) E}}{\text{E + D}} = \frac{(2 \times 10^4)(2.5)}{82.0} = 610 \text{ gal high proof}$$

Film Method

The material for flor sherry is made by fermenting a neutral white grape juice to near dryness. The most common varieties chosen are Palomino or Tokay. They have a neutral flavor and aroma and the desired acidity and pH balance. Other neutral grape varieties can also be used. Thompson Seedless or red varieties should be avoided. Add sufficient tartaric acid to adjust the pH to 3.4 if necessary. A good choice for fermentation is the flor yeast *Torulaspora delbreukii* which is available in dry form. When the fermentation reaches less than 0.2% reducing sugar, fortify the fermented must to 15-15.5% v/v ethanol.

An American oak barrel should be filled 1/2 to 2/3 full with the fortified wine. Place two or three layers of cheesecloth over the bung hole and apply wax around the edges to keep flies out. A film should develop in a week or two. Let it remain undisturbed at 15°C-18°C. (If temperatures get over 25°C, the film can be destroyed.) It is worthwhile to check the alcohol and volatile acidity every month or two. If the alcohol goes down below 14.5% v/v, refortify to 15.5% v/v.

Start a new barrel the following year. Remove part of the first year's product. Add some new wine to the first year's barrel. This encourages continued film growth. Start a new barrel the third and fourth years and you have a solera going. Set up a scheme for transferring a portion of each barrel to an older barrel each year. Remove a portion from the oldest barrel, replace with wine from next oldest, etc. The size of the solera will depend on one's needs. Baker et al. (1952) describe the theory and practice of fractional blending. Commercial production costs are prohibitive for this type of sherry in the USA.

When the wine is taken out from underneath the film, it can either be used at that alcohol level or fortified to 17% v/v ethanol and transferred to a clean barrel for further aging. Some winemakers prefer to mix the flor wine with baked sherry. The latter previously has been fortified to 18-19% v/v and heated to 45-55°C for two months. The wine is then stabilized and held in 50-gallon barrels. The home winemaker can use smaller barrels. In addition, he can remove portions and sweeten to taste.

Submerged Method

Flor sherry has been made commercially by using the submerged culture technique developed by Amerine (1958), Ough and Amerine (1958) and Ough and Amerine (1972). The process is fairly simple. Wine is fermented to dryness and a culture of any fermenting yeast of *Saccharomyces cerevisiae* or a flor yeast such as *Torulaspora delbreuckii* is added. Figure 6-3 shows an example using the latter yeast. The tank is put under pressure of 1 atm and stirred. Commercially, a tank is used of a height so an average pressure in the tank is 1 atm. This would be a tank about 64 ft. high. This is an estimate based on the pressure of a 32-ft. column of water is equal to 1 atm of pressure. The pH of the wine should be adjusted between 3.2-3.4 and the ethanol between 14-15% v/v. A pump arrangement should be set up so the wine is drawn from the top of the tank and comes in the bottom. A portion of the air in the head space of the tank should be drawn in with the wine. The tank should have a small opening at the top so some fresh air gets in to replace the oxygen that is used. This is a relatively small amount. Pumping over should be continuous. The wine and air being circulated should pass through a system to disperse the air into fine bubbles. The holes must be large enough so as not to become clogged by the yeast. The best way to follow the success of the system is to measure the acetaldehyde. Yeast cell population will grow through several log phases with a wine with ample nutrients. Sufficient amino acids for yeast growth must be present. During the successful yeast growth, the acetaldehyde can reach a maximum value of over 1200 mg/L. At this point the yeast die off. Transfer of fresh sherry out and wine in should be made before this level of acetaldehyde is reached. Acetoin also accumulates. There is no change in the 2,3-butanediols and the acetic acid is used up. Ethanol is oxidized to acetaldehyde. When the aldehyde formation stops, so does the acetic acid metabolism but not the acetoin formation (Figure 6-4). A cavitator (a unit that stirs the wine rapidly with aeration) can build up pressure on the blades to cause acetaldehyde formation also (Luthi et al., 1965).

Changes in the drinking habits of the public (lighter table wines and coolers used for aperitif wine rather than sherry) have mini-

FIGURE 6-3. A submerged culture flor sherry fermentation using *Torulaspora delbreuckii* (also called *S. fermenti* and *S. beticus* strain Xerez).

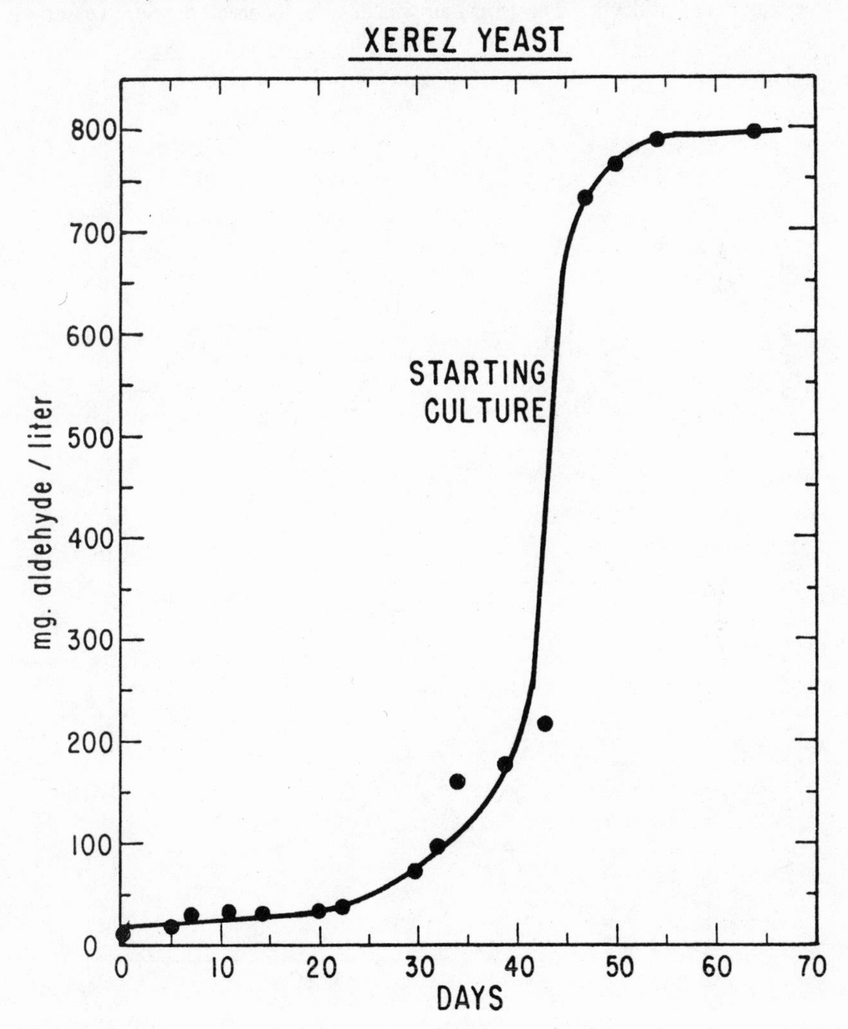

FIGURE 6-4. The effect of pressure on the formation or decomposition of 2,3-butanediol, acetic acid, acetaldehyde, and acetoin. The 0-psi air was done in a cavitator type of device. The 15-psi air was done in the usual pressure vessel.

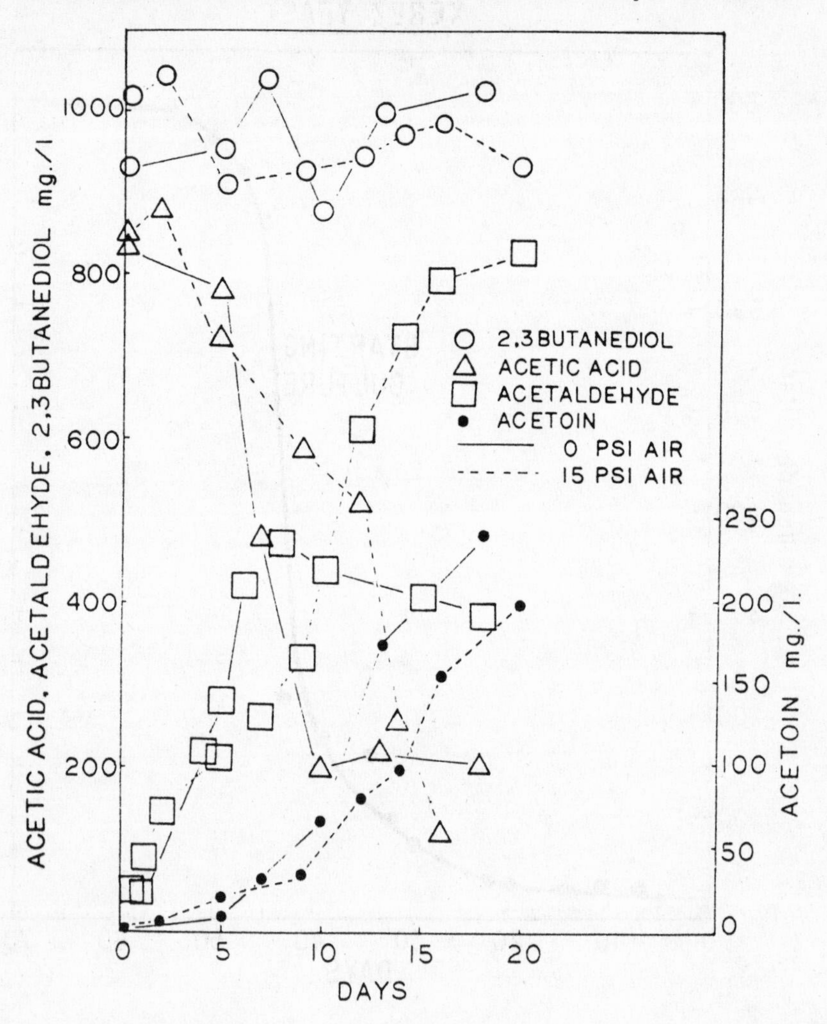

mized the production of this type of wine. A light flor sherry has more resemblance to a table wine than does the usual baked sherry.

CHAMPAGNE OR SPARKLING WINE

Champagne derives its name from the Champagne district in France. Under the rules of the governing O.I.V., even sparkling wines produced in other districts in France cannot be called Champagne. The United States did not sign this agreement and calls many sparkling wines Champagne. Commercial considerations make the nomenclature "sparkling wine" untenable. Any wine made by the direct addition of carbon dioxide is called "carbonated wine."

Champagne in the USA has become more popular in recent years for several reasons. Quality improvement and modest price increases are probably the major reasons. The increases in quality have been because of the increased availability of Chardonnay and Pinot noir grapes grown specifically for this purpose. A strong influence has been seen from French champagne houses setting up business in California. Their practices based on long successful experience have spread to other wineries. Cost control has come from better and less expensive methods of riddling and from strong competition. The product of the charmat process vintners has also shown significant improvement in quality and in competitive pricing.

Primary Fermentation

The very best sparkling wines are made from either the juice of Pinot noir or Chardonnay grapes. The Pinot noir or Pinot Meunier grapes are picked reasonably early, usually 18-20°Brix, and the juice is pressed as described earlier. The primary concern is to have a light juice with as little red pigment as possible. The addition of SO_2 will bleach the juice. This is only temporary and actually acts to stabilize the color pigments that might be lost during fermentation by oxidation and polymerization. Other treatments to remove the color are bentonite and gelatin and silica gel fining. Use of carbon probably removes important flavor components. PVPP can be used. The flavor in the grape is important and no undue harsh treatments should be employed.

Chardonnay should be picked to give no more than 10.5-11.5% alcohol v/v and have a good total acidity of at least 8.0 g tartaric acid/L. If the acidity is too low (pH too high), the juice should be acidified to at least 8.0 g/L of tartaric acid. Other varieties may be used if the two varieties above are not available. Chenin blanc and French Colombard are satisfactory but French Colombard can give some odd aromas if the grapes are harvested too late. Not many other varieties are satisfactory for sparkling wines. When varieties other than Pinot noir, Pinot Meunier or Chardonnay are used, the addition of a small amount of juice from labrusca varieties (Delaware, Catawba, etc.) prior to the second fermentation may be desirable, especially for charmat process sparkling wines. These labrusca varieties add aroma particularly suited to less expensive sparkling wine aroma and flavor made from wines with low varietal aromas. The amounts used are small and the effect should not be overriding.

The juice for sparkling wines is usually fermented cold in the same manner as other wines. Dunsford (1987) investigated the use of malolactic fermentation in the Champagne region of France. She found that most houses did encourage it. Moët-Chandon stopped an active wine malolactic fermentation in midpoint. Then they held the wine at −2°C. They warmed it up the following year to use as a starter culture with excellent success compared with the usual methods of inoculations. Wines are cellared for about one year, blended and normal stability treatments done and fining and filtering carried out.

Bottle Fermented

For the normal bottle-fermented method, a culture of Champagne yeast *Saccharomyces cerevisiae* race *baynus* or other satisfactory yeast is cultured. It must be able to grow and ferment well at the alcohol concentrations of the wine. Figure 6-5 shows a standard yeast propagation system for champagnes. The yeast should impart desired flavors and be of a flocculating nature. Most winemakers acclimatize the yeast to SO_2 and ethanol by growing at the concen-

FIGURE 6-5. A modern yeast adaption and propagation system for all types of champagne.

trations present in the wine to be used. Monk and Storer (1986) indicated that starter culture preparation was important if the level of inoculation was below 4×10^6 viable cells/ml. Above that level of inocula, the pretreatments had no effect. Below 1×10^6 cells/ml, fermentation lag was 25% of the total fermentation time for starters prepared either anaerobically or aerobically in wine. In juice, the lag was 10-15%. The use of anaerobically prepared cultures resulted in faster times to completion.

It is usually customary to put a small amount of bentonite and other adjuncts in with the yeast inoculum and the sugar. This not only helps with the fermentation but is a considerable aid in the riddling process. The inoculum can be cooled to stop the fermentation, or lightly centrifuged, to separate off about 3/4 of the supernatant liquid. Use the rest as the inoculum.

Moët-Chandon has successfully experimented with the use of immobilized champagne yeast for the secondary fermentation. Fumi et al. (1988) found that calcium alginate gel, used to immobilize champagne yeast, did not cause much change in the fermented product. There was a slight increase in calcium in the sparkling wine, time to ferment was unchanged, and the use did not affect the clarity of the wine. Slight differences in aroma, composition and amino acids were noted, however. Labor savings by the use of immobilized cells are significant.

The wine with the proper amount of invert sugar added is inoculated. Sucrose may be used but it must first be heated with a small amount of acid present to invert it. The sugar is dissolved in water, acid added (1-2% wt/vol citric acid) and brought to a boil. It is then cooled before use and brought to volume. The amounts to add to attain the pressure desired by the yeast fermentation of the sugar are given in Table 6-2.

Assume 1.0 g/L reducing sugar is residual. The amount of dry sugar lb/1000 gal to give 6 atm at 10°C is about 200 lbs at 1 g/L residual reducing sugar. To convert these values to dry wt. sugar, for example, for 100 gal with 4.0 g/L residual sugar to get 6 atm pressure.

$$\frac{159.0}{181.7} \times 200 = 175 \text{ lbs dry sugar/1000 gal.}$$

Mix the appropriate amounts of sugar and wine together. Then add approximately 2% (by volume) of the concentrated yeast culture or the immobilized cells. All three ingredients should be at the same temperature (10-15°C) before mixing. While still mixing, the prepared wine is put into champagne bottles. Be sure that the bottles will be able to withstand the pressure of champagne fermentation. The explosion of the bottle during disgorging later is not uncommon when using new champagne bottles. Most of the bottles used to ferment sparkling wines have a lip for a crown cap which is the usual means of closure before disgorging. A standard crown bottle capper is used to put on these caps.

The bottle should be filled leaving about 1 inch (2.5 cm) of air headspace after capping. The filled and capped bottles are laid on their sides and allowed to ferment at 12°C ± 2°C. After bottling, safety glasses and gloves should always be used when around or handling the bottles. After a week or so, the wine should be cloudy

Table 6-2. Amount of 50% wt/vol Sugar Solution Needed for CO_2 Production for Champagne.

		Volume of 50% wt/vol invert sugar			
		liters		gallons	
Amount of base wine	Reducing sugar in base wine g/L	5 atm	6 atm	5 atm	6 atm
1000 liters	1.0	40.0	48.0	10.6	12.7
	2.0	38.3	46.0	10.1	12.15
	4.0	35.0	42.0	9.2	11.1
	8.0	28.3	34.0	7.5	9.0
1000 gals	1.0	151.4	181.7	40.0	48.0
	2.0	145.0	174.1	38.8	46.0
	4.0	132.5	159.0	35.0	42.0
	8.0	107.1	128.7	28.3	34.0

and usually will ferment to dryness in several weeks. The yeast settles out on the wall of the bottle. For the best quality sparkling wine, the wine should be kept undisturbed for a year or more.

The use of a plastic bidule or insert into the neck of the bottle and the addition of small amounts of bentonite or silica gel before bottle fermentation are common practice. The bidule will trap the riddled yeast and sediment. This allows a cleaner disgorging. The fining agents allow quicker and cleaner riddling.

The next step is riddling. This is the process of working the yeast down into the neck of the bottle. The first step is to loosen up the yeast. This can be accomplished by the use of a rubber mallet or by suitable shaking devices. If a mallet is used, wrap a heavy towel around the bottle (for safety in case of explosion). Strike the bottom firmly and vigorously with the mallet. Do not strike the bottle any place except the bottom as it may explode. Safety glasses and gloves must be worn. Once the yeast has been loosened from the sides of the bottles, put it into the riddling rack (Figure 6-6). The neck should be down and the bottle at a 45° angle. A mark for reference is made on the rim of the bottom. The bottles are given a 1/4 turn to the right on the first day, 1/4 turn to the left on the second day, another 1/4 turn to the left on the third day, 1/4 turn to the right on the fourth day, and 1/4 turn to the right on the fifth day. This sequence must be repeated until the yeast sediment is worked down onto the cap. This will take several weeks.

New automatic riddling bins have taken over most of the riddling chores in most champagne houses. These are made so they can be rotated in several planes. They can hold hundreds or even thousands of bottles. They can be vibrated at a desired frequency and intensity. They are motor driven and can be programmed to give optimal results. It takes about 7 days to accomplish what it takes about two weeks to do with the hand riddling. One winery is automated even more. The champagne bottles are cased when the inocula is put in. Riddling takes place in the cases on an automatic track system. Very little handling is required. Robotics use is prevalent.

The amount of time the bottled champagne stays on the yeast is a question of economics as well as quality. There are changes that occur in the yeast and wine as time goes on. The ester content

FIGURE 6-6. An A-frame riddling rack. The champagne bottle is marked on the base. It is turned a fraction of a turn each day.

changes, some esters form and some disappear. The yeast begins to autolyze. The yeast has several protease enzymes that can be liberated. They will cause small amounts of peptides, mainly those containing alanine and arginine, to be hydrolysed from the cell according to Lurton and Guerreau (1988). Table 6-3 lists the proteases and some information about them.

Feuillat et al. (1988) showed that colloids formed by yeast in the first eight months of aging of bottle-fermented champagne can aid in bubble formation and aroma release. The colloid is released from the cell walls. It is a glycoprotein, 75-87% sugars and 3-4% nitrogen. The sugars were primarily mannose, 43%; glucose, 31%; inositol, 15%; with galactose, arabinose, fucose and rhamnose making up the rest.

Other considerations that affect the style of the champagne are the temperature of fermentation and storage. Champagne fermented warm (>20°C) has a harsher character and seems to be less fruity. Warmer storage conditions (on the yeast) give a mellower, more complex, flavor. The same wine stored cold (11°C) has a fruitier

Table 6-3. Characteristics of the Principle Proteases of Saccharomyces cerevisiae.

Name	Action	Inhibitor	pH optimum
Protease A	acid endopeptidase	pepstatine	3
Protease B	serine sulfhydryl endo	PMSF[a], $HgCl_2$	7
Carboxypeptidase Y	serine exopeptidase	PMSF[a], $HgCl_2$	6
Carboxypeptidase S	metallo exopeptidase	EDTA[b], PMSF[a]	-
Aminopeptidase I	metallo exopeptidase	EDTA[b]	8
Aminopeptidase II	metallo exopeptidase	$EDTA_b$, $HgCl_2$	7
Aminopeptidase Co	metallo exopeptidase	$EDTA_b$, $ZnCl_2$	8

Lurton and Guerreau (1988)

[a]Phenylmethylsulfonylfluorine

[b]Ethylenediamine tetraacetic acid

aroma but is less mature and is simpler. Once the storage period is ended the champagne is disgorged. This is usually accomplished by either of two methods. In the regular disgorging procedure the bottle, after riddling, is inverted and carried through a brine bath. The bath must be cold enough to free a plug of frozen wine in the neck of the bottle about 1 1/2 inches long. The bottles are chilled to 0°C before this process starts. The cap is then removed and the plug of ice scrapes the yeast from the neck of the bottle as it is forced out. Only a small amount of liquid is lost as the bottle is turned upright immediately after the plug is gone. The bottle is refilled with a dosage containing some sugar and a little SO_2. The bottle is then corked and ready for labelling. These operations can be automated.

Transfer Process

The second method called transfer process is the same until the wine is ready for disgorging. The wine is completely removed from the bottles. This disgorged wine is pooled and cold stabilized, while contained under pressure. The wine is then filtered and put back into the bottles for corking and labelling. The dosage is either added before the bottling or added to each bottle as they go to the filler. All this is done while the wine is very cold and mostly under counter pressure.

Invert sugar is usually used for champagne sugar additions. This is at the tirage (yeast-sugar addition to the wine) and at the dosage (addition of sugar for sweetening to champagne). The sugar solution is 50g/100 ml. The amounts used are variable depending on desired sugar content.

Champagne corks are made of several pieces. The bottom piece that comes in contact with the wine is usually of the highest quality. The pieces are designed to expand to fit the neck of the bottle. Special water-resistant, odorless and tasteless glue is used for putting the pieces together. The champagne cork is just a large, straight-pieced cork before it goes into the bottle. After it has been in the bottle for some time it develops the typical mushroom shape. When the cork is first put in it is very difficult to remove. As the time passes, the cork relaxes and can be pulled out much easier. The metal hood is put on to prevent heat changes from moving the cork out later. Agglomerate corks are in common use.

It is probably best to put an explanation on the bottles on how to remove the cork safely. Eye damage can result from careless removal or improper cooling of the champagne before opening.

Bulk Fermented (Charmat Process)

Processing champagne by the bulk process is much less labor-intensive than is the bottle method. The size of the tank to be used is only limited by the engineers' ability to make it structurally safe. Tanks as large as 30,000 gal have been used. The process is very similar as far as the fermentation goes. Refer to Figure 6-7. Tank A is filled through valve 3 with valve 5 open. All valves are closed after the desired wine, sugar and yeast are in the tank. The temperature can be from 12°-22°C for the fermentation. When the ferment reaches the proper sugar content or is dry, the wine is chilled to

FIGURE 6-7. Schematic of bulk fermentation tanks for sparkling wine. 1. Cooling jackets, 2. Thermometers, 3. and 6. Draining valves, 4. and 7. Racking valves, 5. and 8. Barometric tubes, and 9. Pressure gauges.

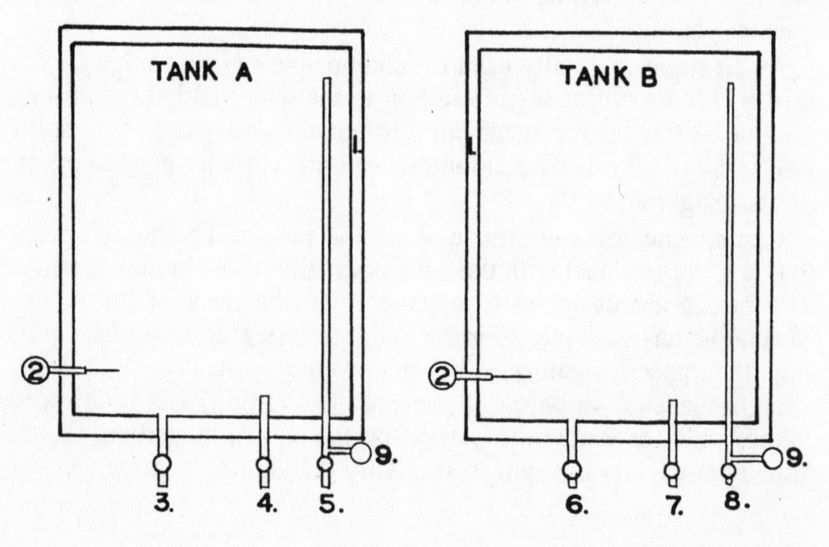

−4°C (24°F). The sugar is adjusted and the wine stabilized for tartrates and the yeast settled. After about a week to 10 days, the wine is filtered from valve 4 of tank A through a plate and frame filter into tank B through valve 6. Tank B should first be chilled to −4°C (24°F) and a counter pressure of N_2 applied slightly higher than the pressure indicated in tank A. CO_2 can be used for this step if the CO_2 does not increase the original pressure. All other valves except 4 and 6 should be closed during the transfer and filtration. The wine can then be bottled cold from tank B.

Once it is in the bottle, a polyethylene closure is usually used as a stopper. These closures are only adequate for about six months to one year. They are permeable to oxygen and allow the wine to oxidize slowly. Some producers try to have their champagne with polyethylene closures slightly reduced with a trace of H_2S at bottling. This will cause some delay in the oxidation. The use of part plastic or ground cork to simulate a normal cork closure has been used with some success. When the bottles are closed on the bottling line, any bottles from which the closure pops out should be recycled. If it would pop out when the wine was cold (colder than if it was being kept in a home refrigerator), then it could be a potential problem. The wine hoods are put on and the bottles are warmed up before labelling to avoid sweating problems causing less-than-perfect labelling.

Once the pressure reaches a certain point in the fermentation, the yeast will no longer grow. They will not be inhibited by the pressure as far as fermentation goes until the pressure is over 100 psi or about 7 atm. Growth ceases at 30 psi, give or take 5 psi. Champagne yeast will not regrow normally in the bottle as long as the pressure is above this value. Addition of SO_2 to the wine will bleach the color. Bacteria, which require a great deal more pressure to stop growth, will be inhibited by SO_2. In addition, SO_2 can have a negative sensory effect on the aroma when the wine is served.

Most continuous champagne systems, such as the one proposed and in use by the Russians, are not truly continuous. They grow yeast on molasses continually and add the yeast into the wine going in. They use CO_2 gas to maintain the pressure desired. The yeast flows through the system but gives very little to it as it is filtered out at the end before bottling. Since the yeast is not acclimatized to the alcohol of the wine and it has a relatively short residence time in the system, it probably contributes more from shock autolysis than actual fermentation.

It is assumed that great changes take place in the amino acid content of the bottle-fermented and aged champagne compared to the charmat or bulk process. In reality, this is not true. Any quality changes in aroma or flavor have to be attributed to other chemical reactions. Colograndе et al. (1984) measured the amino acids residual in commercial sparkling wines. He tested the charmat process and the bottle-fermented process. The number of different samples was 10 and 23, respectively. There was very little difference in the average values determined between the two types of champagne. The values are listed in Table 6-4. (The method of analysis was dansylation directly after the sample was dried.)

Unless the proteins and peptides of the yeast are broken down by protease activity to form amino acids, this result is expected. The free amino acid pools within the yeast cells are hardly large enough to be measured when released into the wine.

The key elements in making superior champagne are: (1) using the proper varieties picked at the proper maturity, (2) good, sound, winemaking practices during the primary fermentations, clarifications and stabilizations, (3) proper blending to get the desired wine composition, (4) choice and preparation of yeast for secondary fermentation, (5) control of fermentation and aging temperatures, and (6) aging of the product.

There are choices to make based on the quality possibilities weighed against the economic realities. If the company's product is bulk champagne, purchasing Chardonnay or Pinot noir grapes is unwise. The product's value as a bulk champagne could not command a sufficient price for the operation to be economically viable. Conversely, the use of the cheaper varieties of grapes for bottle-fermented champagne lowers the quality and degrades the label.

Carbonated Wines

Carbon dioxide is quite soluble in water and in wine. For example, at 0°C 1.71 volumes of CO_2 (at standard temperature and pressure) will be dissolved and the gauge pressure will be 0 psi. Under

Table 6-4. Commercial Sparkling Wine Amino Acid Analysis mg/L.

	Fermentation type			Fermentation type	
Amino acid	Tank	Bottle	Amino Acid	Tank	Bottle
Asparagine	37	30	Isoleucine	83	46
Arginine	132	138	Leucine	7.8	9.0
Aspartate+			Ornithine	18	9
Glutamate	34	34	Lysine	21	29
Threonine	70	61	Tyrosine	15	12
Glycine	22	20	Ammonia	14	19
Alanine	71	77			
Proline	690	470			
Valine	10	9.9	Total amino acid	182	184
Methionine	16	23			
Tryptophane	6.9	13	Total nitrogen	263	280
Phenylalanine	13	15			

Colograndе et al. (1984)

The authors could not statistically separate the two types of champagne by the amounts of amino acids.

100 psig pressure, 13.4 volumes of CO_2 will be dissolved. As the temperature is raised, the volumes of gas that can be dissolved decreases. At 20°C at 0 psig, the volume dissolved is 0.88 and at 100 psig it is 6.8.

"Carbonated" wines are those that are saturated with CO_2 at some low temperature and under pressure. When they are raised to serving or storage temperatures, they will have sufficient CO_2 to resemble sparkling wines, but not so much as to explode the bottle or be unsafe to handle.

This type of wine does pay an elevated tax, only slightly less than sparkling wine (champagne) tax. Very little is produced in California, but in other parts of the world it is produced in significant amounts. The obvious advantage is rapid production and low costs. As stated in Amerine et al. (1980), if the tax were more reasonable and the term "carbonated" on the label not mandatory, it could be a desirable product. In the United States, any wine over 0.392 g/100 ml of CO_2 is either considered carbonated wine or, if the CO_2 is from the yeast fermentation only, then champagne and taxed as such.

Chapter 7

Aging, Bottling and Storage

Once the wine has been fermented and stabilized, the next steps are aging, bottling and storage. All wines benefit from aging, some more, some less. Aging can be done in tanks, bottles or barrel. Aging allows slow chemical changes to occur in the wine which cause the wine to mellow and improve in quality. Choice of barrel wood can also affect the taste and aroma of wine. White wines are aged differently and usually for less time than red wines. Bottling of the wines at the proper time is important. Over-aging of wines in the barrel or bottle is harmful. Storage conditions must be properly maintained if the bottled wine is to continue to mature and improve.

TAINTED WINES

There is nothing worse than ruining a good wine by contamination. Strauss et al. (1985) listed several problems that they had encountered involving tainted wines. They divided the problems into three major categories: (1) off-flavors from defective additives or packaging material, (2) ones derived from spillage, and (3) those resulting from unsuitable or defective equipment. In the first category were impure gasses, contaminated corks and contaminated water used to dilute spirits. In the second category were direct oil leaks from equipment, spills on or near tank linings which were absorbed into the wax and leaking refrigerants. In the last category were unsuitable storage tanks from which off-odors and tastes could be extracted and defective paints used to line tanks. The investigation of these types of contamination is difficult without sophisticated equipment. It is best to avoid these problems. This can be done by careful selection and use of the proper cleaning compound, use of certified wine tolerant materials and care and common sense.

STORAGE AND AGING

White Wines

White wines begin to "age" as soon as the yeast stops fermenting. White wines of the fresh, light (10-12% ethanol) and fruity type should have a minimum of aging if the aromas are purely from yeast fermentation. The maturation process in these types of wines is relatively minor. By the time they are clarified and stabilized, they are near ready to drink. A few months in the bottle held at a low storage temperature is all that is required. These wines depend primarily on the aroma of the esters developed during yeast fermentation. These esters disappear rather rapidly with time (the warmer the storage temperature, the more rapidly they disappear). These wines are usually good up to two years or less before they lose their appeal.

An alternate style is white wine that had more insoluble solids contact during fermentation and is more complex. This wine is usually not strongly fresh and fruity in character but has other wine flavors besides the acetate esters. This wine will benefit by aging. This is usually done in larger casks or tanks. This wine, after aging, can benefit with a bit of blending with a fresh-fruity wine to add even more complexity.

Some white wines are barrel aged. This usually is for wines that are going to be sold for high prices. They must have sufficient body and character so the oak flavor does not overpower the wine. Certain Chardonnay wines are examples; those that have some skin contact, those blended with a percentage that have had a malolactic fermentation, and some that have had prolonged yeast contact.

In Burgundy and the Loire, white wines can be kept on the lees up to two years. This treatment is "sur lie." The yeast is resuspended occasionally the first few months. This does slightly enrich the nitrogen content of the wines due to autolysis. The main benefit is that a light color is maintained by the yeast adsorbing the oxidized brown material. Comparisons made by Feuillat (1987) found that the control wine was preferred to the lees contact wine after

FIGURE 7-1. An automated barrel resurfacing machine.

bottling and six months' bottle age. After 18 months of bottle aging, the lees sample was better appreciated. Usually the barrel time is much less than compared to red wines. The number of barrels needed, therefore, to age the wine is relatively small if used efficiently. Barrels for white wines of this type are usually replaced every three or four years. In some instances, they are used only one year, then reused for red wines. They can be used for another four years with red wines. They can be resurfaced inside and used for another several years. The time a barrel is kept depends somewhat on how much oak flavor is desired. Wines will age very well in old barrels that impart no oak flavor. Figure 7-1 shows an automated barrel resurfacing machine.

Pocock et al. (1984) found ellagic acid as a deposit in bottle-aged white table wines. The wines had been treated with oak shavings or chips just before bottling. The hydrolysis of the ellagitannins gave glucose, gallic acid and the diphenyl compound which will condense to ellagic acid. (See p. 218.)

Ellagitannin → + Glucose + Gallic Acid

$-2H_2O$

Ellagic Acid

The aging of white wines is done with the use of SO_2. Wines made without any SO_2 at the crusher or during fermentation are almost always protected by SO_2 from oxidation during their aging and storage period.

The amount of SO_2 used will vary with the winemaker and situation. Wines held in barrels require a larger addition of SO_2 to protect them from bacterial activity. Wines held in stainless steel tanks with oxygen excluded have a lower requirement. The actual amount of free SO_2 required will also depend on the pH; the lower the pH the less required (see Chapter 10). The usual amount of free SO_2 is between 20 to 30 mg/L. Air must be excluded from the white wines. This is done by using non-aerating pumps or gravity to move the wines and use of inert gas blanketing when necessary.

White Riesling wines age differently than most of the varieties. Sometimes they take on a kerosene-like aroma after a year or two in the bottle. This aroma (Simpson and Miller, 1983) is due to vitispirane and 1,1,6-trimethyl-1,2-dihydronaphthalene (TDN). Heating will speed the formation of TDN in either the juice or wine and low pH encourages the formation. Emerald Riesling was also shown by these authors to have a significant amount of these two compounds. While this aged bouquet can be appreciated, it is preferred that it be a subtle contribution to the overall aroma and not the predominant odor.

Rosé and Blush Wines

Rosé or blush wines are usually treated more like fresh-fruity white wines. In these wines, color, body, sweetness and freshness are all critical. First, the color must be bright and in a certain red range – not too purple, brown or orange. It should have a mouth-filling flavor. Certain varieties lend themselves to this better than others. Zinfandel and Grenache both make excellent wines of this type. To maintain Grenache's color stability, the addition of SO_2 is required. The level of sweetness should be such as to balance the wine and cover any harshness or bitterness that may be present. It should not be sweet enough to cause one to think sweet. Maintaining color is a big problem for the winemaker. He/she should protect the wine carefully from oxygen and maintain the SO_2 at the proper level. Also, if the color becomes less desirable, he/she should have some material to blend with to adjust the color. This material should be a similar wine made with more skin contact and aged so the pigment is complexed with other phenols.

Red Wines

Red wine aging takes longer than white, rosé or blush wines. The reason is the phenols present that are extracted from the skins during fermentation and pressing. Most of the varietal and other desirable aromas are associated with the skins.

Red wines should be through their malolactic fermentation before aging. Red wine are usually pressed near dryness. When dry, the wine is kept on the yeast lees at 20°-25°C. It is inoculated with the desired malolactic. Once the malolactic fermentation is complete, the wine is fined, cold-stabilized and filtered. The SO_2 is adjusted and the wine is ready for aging. If the wine is not to have a malolactic fermentation, the wine is quickly taken off the lees. The SO_2 is adjusted to desired level, the wine is cold-stabilized and filtered. It is then maintained with sufficient free SO_2 to inhibit malolactic bacteria. The amount of free SO_2 is dependent on the SO_2-binding compounds in the wine. The active form (molecular SO_2) is dependent on pH and free SO_2. These wines are usually kept in stainless steel tanks for aging.

Traditionally, in Bordeaux, the red wines are aged for two years in the barrel. For the first six months they are aged bung up. This allows the bung to dry out and for more oxygen to get to the wine. Then the barrel is turned on its side so the bung is wet and less air transmitted into the wine. The wine in Bordeaux is not stabilized or filtered before barrelling. The barrels are racked before January. They are racked every three or four months for the first year. This racking also gives more oxygen to the wine. The second year the wines are fined with egg white (5 or 6 egg whites per barrel). The bottling time is usually determined by an experienced taster-winemaker.

When red wines are aging in barrels (or otherwise), care must be taken to check for spoilage organisms and changes in color, aroma or flavor. Errors in too little SO_2 can cause the spoilage of a wine by acetobacter or yeast. Most spoilage of a wine by bacteria is easily controlled by SO_2 at the proper levels and by exclusion of unnecessary oxygen. It is a very good policy to sample the barrels or at least smell them each time they are filled. If there is any question, the wine should be checked for acetic acid and SO_2. Records should be kept to determine changes so a wine can be "fixed" before it is spoiled. The loss of even one 50 gallon barrel of Cabernet Sauvignon or other varietal wine due to spoilage is expensive. Wine can spoil in a full stainless steel tank if the SO_2 level is not appropriate.

The barrel of wine should take up 30 ml/L of oxygen per year (Peynaud, 1984). This can occur either through the bung, by racking and refilling or absorption through the wood. Evaporation alone can account for 10% loss in the barrel-aged wine over a two-year period. This can vary somewhat with temperature and humidity. Up to 200 mg/L of phenols can be absorbed from the wood by the wine during aging in a new barrel.

The position a barrel is stored in varies considerably from winery to winery. The Bordeaux method was described above. Other ways are: upright with airtight bung, on side with tight bung from the start, setting on end with a tight bung on top. Storage systems vary. There are pyramids, wood runners—barrels rolled in and out (Figure 7-2), metal stack pallets (Figure 7-3), wood pallets (barrels upright) (Figure 7-4), and special aluminum racks holding individual barrels. Some wineries may only check and fill barrels every six months or so. Others do it monthly or even oftener. The object is to age the wine to a desired level of maturity without it being microbiologically spoiled or changed in a negative manner. There are advantages and disadvantages to each system. The pyramid stack is very space efficient. It is difficult to fill barrels and time-consuming to handle them. Metal pallets are easy to handle and are as space efficient.

A good part of aging a quality red wine is knowing when and how much and with what to get the desired flavor, texture and balance in the wine. The technical aspects of the preparation and addition of the fining materials are given in Chapter 4. This is the easy part. The difficult part is only learned by tasting experience. The treatment results cannot be described effectively with words with-

FIGURE 7-2. Racks for barrels. Each can be filled or emptied in place and rolled in and out for cleaning.

FIGURE 7-3. Metal stack pallets for efficient storage. Must be handled to fill or empty.

FIGURE 7-4. Wood storage pallets. Efficient for space but requires moving for filling and emptying.

out the trained sensory experience. The higher the phenolic content in the wine, the more it will benefit from fining. Tasting and evaluating and keeping full records of your impressions during the wine's aging is essential. Some wine types require relatively little aging. Zinfandel grown in the cooler areas has a strong raspberry-like character. This will fade and disappear in several years. To get the full benefit of this very pleasant style of wine, it is aged for only a short time. It is bottled after a year at the latest. If it is kept in the barrel for longer times, this character is lost and the wine becomes more ordinary without a specific varietal character. Zinfandels with more alcohol may require longer wood aging.

The quality of the wine should be well considered before putting a wine into an expensive barrel. If it is an ordinary wine, it will still be an ordinary wine after barrel aging. The return may not justify the expense of barrel aging.

There is the question of what temperature table wine should be stored at. Experience indicates that 10-15°C may be the best range. Perhaps red wines a little on the high side of this range and whites on the lower side. A case can be made for higher or lower temperatures. At higher temperature, wines mature faster and can get on the market sooner. However, some adverse flavor changes occur that don't occur so readily at cooler storage. The pros and cons have to be considered – small quality differences vs. extra costs in refrigeration, and extended time without return on investment.

Barrels

Barrel aging of red wines is not standardized. American oak, German oak, French oak (of several kinds) and Yugoslavian oak are all imported. They are used more or less at random throughout the wine industry of the world. The variability between the growth of the trees of the same species between one location and another can cause a significant difference in the wood. The wood-aging process can vary significantly from barrel maker to barrel maker.

Wines in two different sorts of oak barrels, Nievre and Allier, were compared with an equal volume of wine in a stainless steel container over a period of six months by Piergiovanni et al. (1988). While it is difficult to draw any definite results without barrel replication from many manufacturers, they did see some definite changes. Toasting broke down the lignins and caused more rapid migration of the non-flavonoid phenolics. This has little effect on the taste or aroma of the wine. The oxidative changes that occurred from the air penetrating the staves caused the greatest changes in aroma, color and taste. Color density dropped in the barrels. The 420/520 ratio (more brown) increased as expected compared to the wine in stainless steel tanks. Sensory differences in wine from the different barrels could be significantly separated by triangle tastings after three months. It is extremely difficult to draw conclusions based on such comparisons. Changes in composition and taste were noted due to differences in the oak density, thickness of the staves, degree of toasting, storage bung, and barrel making. The differences in wine quality associated with different forest areas of the same species is subject to question. Perhaps it contributes more to the selling mystique than to the real wine quality in general.

The actual making of the barrel is a well-developed art in France. Cordier (1987) has described the process. The wood is carefully selected from the various areas (center of France, Allier, Nievre, Vosges, Argonne and Limousin). Only these areas give top quality wood in France. The wood is inspected for defects, sorted and taken to be split. The wood is split radially from the center. The savings in splitting over tangential sawing are due to less loss because of saw blade cut. The wood is then aged in the open air to decrease the tannins and to dry the wood out. When the moisture content reaches 15%, they are worked into staves. The staves are put into a jig. The heads are put together using reeds and oak pegs. The barrel staves are heated to bend into shape. During the shaping and putting on the hoops (steel bands), a certain amount of toasting is done. After the shaping is done, the inside of the barrel is heated. The degree of heating determines the level of toasting. The heavier the toasting, the more compounds will be drawn from the wood. There are four choices (none, light, medium and heavy). The heads are put on. The barrel hoops tightened, leaks are checked for. Any wood borer leaks are sealed with an oak peg (spile). If a barrel dries out before use, then it may leak. The cooper can fix this usually by pegs and wedges and hoop tightening.

During the toasting there are increases in vanillin, phenolic aldehydes and acids. Also, phenolic alcohols increase. French oak has more extractives than American oak.

New barrels should be first filled with water and inspected for leaks. If any show up, the barrels should be kept full of water for 24 hours. If the leaks do not stop by the wood swelling up, then they should be returned to the seller for exchange or repair. All new barrels should carefully be inspected for wood borers. These are evident by holes in the staves about the size of match-sticks or slightly smaller. If this is noted, return the barrels or do not purchase them. Borers usually occur in wood kept outside and near old wood, but seldom inside the winery.

When you fill a tight barrel and handle it properly, no problems will usually occur. When you empty the barrel, rinse it out with water and refill it with wine if possible. If this is not possible, then a sulfur wick should be burned in the barrel and it lightly bunged. More commonly now it is filled with a small amount of SO_2 solution (1000 mg/L). One of these treatments should be repeated as often as necessary to maintain sterility in the barrel. Once mold grows inside a barrel, it will never be a really good container for aging wine. The smell of the mold will penetrate the wood and is very difficult to remove without extracting all the oak flavor.

If there is only a very light mold infection (by smelling the empty barrel), then treatment with a 1% solution of caustic soda is suggested. This is followed by a thorough rinse with a 5% solution of citric acid. Lastly rinse with water and burn a sulfur wick in the barrel. It is usually carelessness when this treatment is necessary. Its need reflects on the winemaker's ability.

Some of the main volatile compounds which increase in the wines upon oak aging are: ethyl lactate, furfural, guaiacol, 4-methyl guaiacol, eugenol, 2-phenethanol, methyl-5-furfural, furfuryl alcohol, and β-methyl-α-octolactone. These are also associated with wines fermented in oak barrels. The holding on the lees in the barrel reduces aldehydes to alcohols and changes the character of the wine. This allows differentiation between wines fermented in barrels and those fermented in stainless steel and then aged in barrels according to Marsal et al. (1988). For an extensive review of wood and its effect on alcoholic beverages see Maga (1989).

Oak Shavings

The use of wood shavings to give wines an oak flavor is seldom employed because of off-flavors and bitterness caused by this treatment. If they are used, Buchanan (1985) suggests that the use of oak wood shavings on white table wines be done at the crusher. Some of the advantages suggested are: (1) better press yields, (2) less oxidation, (3) ellagic acid instability is reduced, (4) less bentonite needed for protein stability, and (5) improved settling. Some disadvantages are: (1) masking of varietal flavors, and (2) blockages of pumps and lines if not well dispersed. Improvement of the color stability of red wines was noted.

Oxygen in Barrels

The presence of oxygen during the storage of red wine in oak barrels causes changes. According to Glories (1987), oxidation of the anthocyanins causes loss of color and tannin. Polymerization also occurs, causing brown color and oxidation of ethanol to acetaldehyde which reacts to bind anthocyanins to tannins. This latter compound is stabler and has more color than the anthocyanin. Eventually these also polymerize and precipitate.

Oxygen Uptake

The rate of uptake of oxygen varies from wine to wine. The reactions of the oxygen with the various components that change are sensitive to temperature. Data in Figure 7-5 show an example of how rapidly a wine saturated with oxygen will react and the oxygen

FIGURE 7-5. Time for the oxygen to disappear from a wine saturated with oxygen as a function of temperature. Can vary with wines.

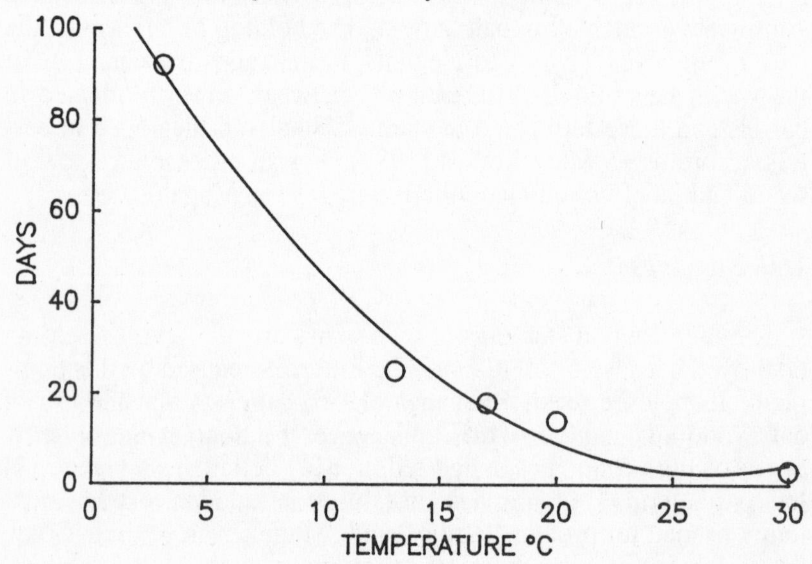

disappears. As long as sufficient SO_2 is present to scavenge the highly reactive oxidizing molecules, the wine is protected from drastic changes.

The uptake of oxygen from air on the undisturbed surface of a wine is about 200 $mg/hr/m^2$ (Müller-Spath, 1973). Thus, the surface-to-volume area of a tank and the exposure time become extremely important when considering oxygen pick up. An example to consider is a tank of 10 gal, 0.3 m diameter (two surfaces), that takes one hour to filter. The calculated amount of O_2 would be about 0.75 mg/L. For a larger operation, it is not that different. A tank with a diameter of 4 meters and 50,000 liter capacity, with 8-hour filtering time:

$$\frac{\pi \times 2^2 \times 8 \times 200 \times 2}{5 \times 10^4} = 0.80 \text{ mg/L}$$

Seldom do you have an undisturbed surface, so these are very conservative estimates. If you halve the diameter of the tank, you decrease the amount of uptake by a factor of 4 with the same wine volume. For the 4 m diameter tank exposed to air for 1 week, the amount absorbed would be 8.44 mg/L, or about saturation level.

The amount of SO_2 required to react with this amount of O_2 is 4×8.44 or about 34 mg/L. SO_2 is not 100% efficient in scavenging the oxidative molecules formed.

Oxygen Removal

In all instances of white table wine storage, the less oxygen that is present the better the product will be. Prevention of oxygen contamination is far superior to removal. In most instances, the quality of the wine is reduced by the removal treatment. The removal of unwanted oxygen from wines can be achieved by sparging with nitrogen. The efficiency of the process will depend primarily on the bubble size of nitrogen used. The smaller the bubble, the more surface to volume, and therefore, the more rapidly the bubble will come to equilibrium. It is better to use a small volume of gas and repeat the sparging if necessary rather than use an excess volume in one sparging. Carbon dioxide can be used but is less effective and saturates the wine with CO_2 which, with table wines, is undesirable. Counter current systems are most effective. In-line systems such as shown in Figure 7-6 are used more often. They are simple and easy to handle.

Inert Gas

Holding wines in partially filled tanks leads to spoiled wines unless the oxygen is removed from the head space. This can be accomplished in several ways. Add dry ice to the tank and allow it to evaporate and replace air (CO_2 is heavier than O_2 or N_2), then close the tank. Fill the tank with CO_2. Pump the wine into the tank from the bottom. Replace the head space over the wine by blowing in CO_2 gas or a mixture of CO_2 and nitrogen. Wilson (1985) found the dry ice method quickly removes the O_2 from the surface of the wine. Unless enough is added to remove the O_2 completely from the tank and the tank sealed, the treatment is relatively ineffective. Diffusion of the remaining oxygen, and that entering an unsealed tank, will soon reach the wine surface. Maintaining a slight but constant CO_2 or N_2 pressure on the head space is a good method to prevent oxidation.

The use of gas mixture (CO_2/N_2) is desirable. This is especially true for white table wines. For a mellow, fully developed flavor in the wine, a certain amount of CO_2 is desirable (Ribéreau-Gayon and Lonvaud-Funel, 1976), and (Lonvaud-Funel and Ribéreau-Gayon, 1977). This usually is less than what would cause significant bubble formation when poured into a glass. If the carbon dioxide levels are

FIGURE 7-6. Schematic diagram of an inline oxygen sparger.

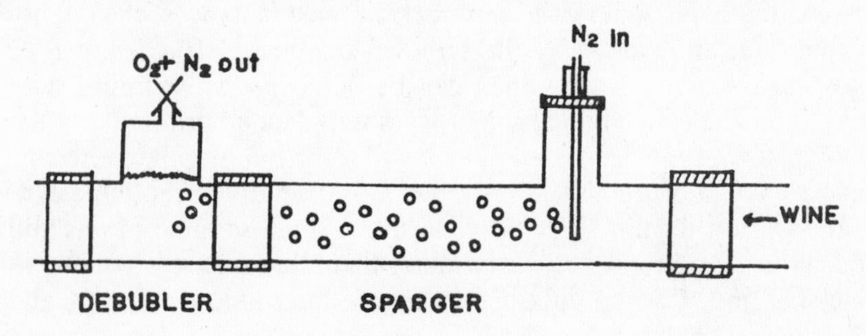

0.2 g/L or less, the wines appear flat; 0.4-0.6 g/L, wines taste developed. Above 0.7 g/L, the wines have a prickly sensation to the tongue. Above 1.0 g/L, the wines appear crisp and some bubbles appear in the glass. At 1.7 g/L, the wine is fully saturated. Each wine may have to be treated slightly differently to reach the desired CO_2 level. Red wines usually need less CO_2 than whites for optimum flavor.

The gas mixtures used can be made up in proportion to the amount of CO_2 desired. If one wants 50% of saturation of CO_2 in the wine (0.85 g/L), then the gas mixture should be 50:50 CO_2/N_2 by weight in the head space. This should also be considered when sparging a wine with N_2 as the CO_2 will be removed as well as the O_2. Temperature affects the total amount of gas absorbed. The colder the temperature, the more gas taken up.

BOTTLING

Before Bottling

Before a wine is bottled it must have complete stability and analytical checks. This involves cold stability test for tartrate stability, a heat test for protein stability and iron and copper analyses. These tests must be run on the final blend of wine. This includes checking for sugar concentration, final SO_2 level, sorbic acid (see Chapter 10 for more information) and the color. Also, if it was not done earlier, the alcohol, total acidity and pH should be determined and then a final sensory examination done of the wine.

The bottled wine should be free of dangerous microorganisms. Wines are filtered through depth filters before they go into the bottling tank. They are clean enough to be put through a membrane filter (0.22 or 0.45 μm) before going into the bottle. Some objections to membrane filtration of Pinot noir wines suggest flavor losses.

If the wine has no sugar and is stable to malolactic bacteria fermentation (i.e., if it has already undergone it), then less attention has to be applied to complete sterility. In all cases, the SO_2 must be kept to appropriate levels.

At Bottling

Oxygen elimination or prevention at the bottling line is essential. The main place where the wine comes into contact with oxygen is at the filler bowl. Gravity feed bottom filling spouts of the filter cause the least oxygen pickup. Figure 7-7 shows a bottling line and Figure 7-8 a bag-in-the-box packaging system. A good rule is that the wine during filling should not increase by more than 1 mg/L of dissolved oxygen. The head space in the bottle, after filling, can be blown with CO_2 or N_2 before capping, closing or corking. It is desirable that the distance between the filler and the corker be short. The faster the line, the more chance of aeration because of turbulence of the wine. Wine bottles or bags should be pulled off the line early in the run and at intervals. They are analyzed for SO_2 ethanol, color, and pH. Samples should be filtered and the filter pad cultured to determine if any microbiological contamination exists (Figure 7-9). This all falls under the domain of quality control. If quality control is not done properly, problem wine can get onto the market. The expense because of consumer rejection can be significant.

Hot Bottling

The pros and cons of hot bottling have been spelled out by Rankine (1984). Table 7-1 lists these. He compares it to cold sterile bottling. The proof of its lack of value probably is seen in that few, if any, California wineries any longer do hot bottling of table wines. However, coolers may be pasteurized in some instances.

Yeast Inhibitors

Many wineries, especially larger wineries, use a yeast inhibitor as a safety measure against possible yeast contamination in sweet table wines. For sorbic acid, the usual level is from 150 to 200 mg/L. This is sufficient to inhibit the growth of yeast in the bottle but not to kill them. A newly approved fungicide dimethyl dicarbonate has recently been approved for use. This compound is an effective agent against yeast found in wines. Its activity is to bind irreversibly the $-NH_2$ or $-NH$ groups on the enzyme active sites. This activity causes cell death.

FIGURE 7-7. A more or less typical bottling line in a medium sized winery.

FIGURE 7-8. A bag-in-the-box operation showing the unfilled bags going into the filling machine and the filled bags loaded in the box.

FIGURE 7-9. A hood with filtered, positive displacement air used for quality control checks of bottled wine for yeast contamination. One bottle of wine is filtered through special systems and the colonies that grow are counted after an appropriate time.

Table 7-1. Advantages and Disadvantages of Hot Bottling.

Advantages	Disadvantages
1. Lower cost-plant	1. Loss of quality, freshness and varietal aromas decline
2. Lower operating expense	2. Some special equipment required for temperature control
3. Better reliability	3. Higher energy costs
4. Some wines benefit from the heating	4. Special heat resistant bottles and closures required
5. Labelling is easier	
6. Holding for sterility checks not needed	
7. Lower levels of general sanitation required	

$$CH_3-O-\overset{\overset{\displaystyle O}{\|}}{C}-O-\overset{\overset{\displaystyle O}{\|}}{C}-O-CH_3 + H_2N-ENZ \rightarrow$$

$$CH_3-O-\overset{\overset{\displaystyle O}{\|}}{C}-\overset{\overset{\displaystyle H}{|}}{N}-ENZ + CO_2 + CH_3OH$$

It also reacts with ethanol to form ethyl methyl carbonate.

$$CH_3-O-\overset{\overset{\displaystyle O}{\|}}{C}-O-C_2H_5$$

This compound can be measured and used to determine how much dimethyl dicarbonate was originally added to the wine. You need to know the ethanol content (Stafford and Ough, 1976). For more information, see Chapter 10.

Smaller wineries that do not wish to add a fungicide or yeast inhibitor may choose to bottle sterile and cold depending on cleanliness. This has been done by many smaller wineries. It is probably more common in Germany. There is a good reason for this. Most of the German sweet table wines are high in sugar. In addition, White Riesling and related cultivar wines have low nitrogen content and a low pH. This enhances the effectiveness of SO_2. All of these conditions add to the difficulty of refermentation.

Microbiology Checks

The report of Neradt (1982) discussed the possible places of yeast contamination during cold sterile bottling. The following points are suggested for checking for sterility: corks, filler inlet and outlet, spouts, bottle mouths, and cork holder bins and standard line samples. This should be done on one day's run over the whole day. Take air samples by exposing agar-filled petri dishes to the air in the bottling room for 3 minutes. The filler was found to be the biggest cause of contamination (48%), with the corker (28%) the next, with bottle sterilizer (10%), bottle mouth (8%) and sterilizing filter (6%) making up the rest. Plate exposure colonies over 100/plate indicated a problem.

Equipment should all be steam sterilizable with no dead ends. The more complicated the system, the greater the chance of an infection. The jaws of the corker are heated and maintained at 80-90°C during bottling.

The filter and line can be sterilized in the middle of a run. The disinfectant should be non-toxic (such as 60 or 70% ethanol with a stabilized oxidizing agent).

Cork Off-Tastes and Smells

Cork is the bark of cork oak trees and comes mainly from trees grown in Spain, Italy and Portugal. It has excellent properties for sealing a wine bottle. The cork must be properly treated and the interior neck of the bottle must be of the proper size. Care should be taken that the corks do not have any chips or striations that could cause leakage. The moisture content of the cork is critical and should be 5-7%. If moisture content is higher, it will seal too rapidly. This causes the air from driving the cork to be trapped under pressure. Later this pressure can cause movement of the cork.

Corked bottles should be left upright 1-3 hours so this excess air pressure can escape. This is especially true if wine is bottled very cold. To check for cork moisture, heat in microwave for 15-20 minutes. Weigh before and after and calculate percent lost as the moisture.

About 13 mm head space is normal and over 20 mm for carbonated wines. Internal pressure greater than 200 kPa will result in leakage.

The proper cork for a bottle is fairly well understood now. The buyer must be sure the cork matches the bottle and the bottles are within specifications. Some problems with oversized, undersized, or not-round corks are not always easily recognized. Corks may not seal properly or may work out. Wine may become oxidized or leak. Bottle irregularities are equally damaging.

Heinzel et al. (1983) offer another solution to sterilizing corks. They suggest the use of peracetic acid. The material they used was 4-5% concentration in water with a cork contact time of 1-2 hours. All the relevant microorganisms were killed. After drying at 60°C for 1 day, the odor of acetic acid was completely removed.

Corky taste in wine can come from several sources – trichloroanisole which has a musty moldy aroma and a threshold of 0.04μg/L, geosmin that has an earthy aroma and 2-methylisoborneol which has a chlorophenolic-like aroma. Chlorine-treated corks can readily form trichlorophenol. This compound can be theoretically methylated by bacteria or molds to form the trichloroanisole. The best way to prevent bad corks from this source is to keep the moisture content low, 5-7%, which prevents microbe and fungal growth.

Tanner et al. (1981) first suggested chloroanisoles were responsible for the cork taint and formed as suggested above. In discussing the source, Maarse et al. (1985) pointed out that oak cork barks treated with chlorophenols were shown to contain chloroanisoles. He favored the theory that they came with the cork from the tree and did not form chemically later. The mold which can methylate chlorophenols includes many species of *Aspergillus* and *Penicillium*. He also noted reports that about 2% of all wine corks were contaminated. Chloroanisoles that Maarse et al. (1985) found in brandy, assumed to come from the cork, are shown in Figure 7-10.

FIGURE 7-10. The chloroanisoles found most frequently in wines and brandies.

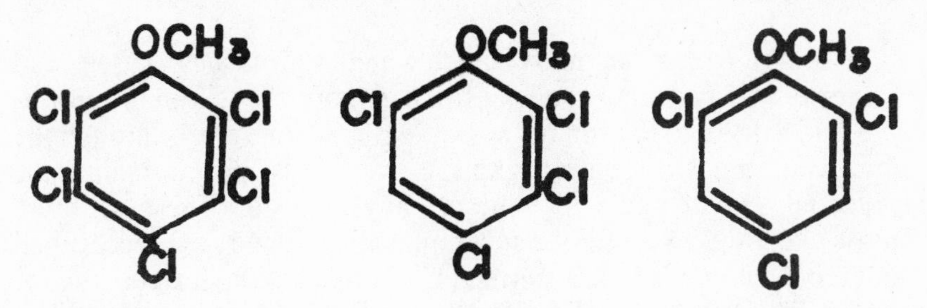

When purchasing corks, you take the risk that they have been contaminated before your possession. Checking for moisture is not sufficient as that could have been modified after contamination. One suggested testing procedure is to place 5 corks selected at random from a sack into a container (suitable laboratory flask). Fill it full with an alcoholic buffer solution (12% ethanol pH 3.3). After 6-8 hrs, evaluate by smell or taste. The difficulty in getting a good sample is obvious. If 2% are contaminated, taking 5 out of 1000 is not good odds in getting one of the 20 contaminated corks. Only a relatively few corks may be contaminated. The test may be favorable but perhaps 2% or more of the bottled wine may end up corky. Corks may come from various sources and be mixed in repackaging. One sack may be good, the next one bad.

To measure cork dust, take 10-20 corks and wash in a bottle by shaking thoroughly with water. Filter the water through a weighed filter paper, dry and weigh filter paper.

Poor coating of the corks can result in paraffin getting into the wines. Some coatings can be soluble in ethanol.

Champagne corks, when sterilized by an ionization treatment, showed no structural changes and were microbiologically sterile. Tests by Rochard et al. (1988) included compression elasticity, angle tension and microbiol infection and moisture content. Even if all tests were satisfactory, this did not prevent corks causing corky tastes from being accepted.

Bach (1988) showed what most winemakers realize: that metal cork- or plastic-lined caps are superior to cork. However, for the public, the cork connotates quality. Closures of other types in quality wines meet severe sales resistance.

Foils

Also, the winemaker must decide whether he wishes to use a lead foil or plastic capsule. The lead foil has more class. It may contribute lead salts onto the rim of the bottle if any wine seeps through the cork and comes in contact with the lead. Most people realize the potential danger and will wipe off any deposits. These salts are most likely the cause for the few high values found in an occasional old wine. Edwards and Amerine (1977) discuss the lead content of wines. Recently lead use for foils has been forbidden in the future.

Labels

Labels are an extremely important part of selling wine. When the product is seen in direct comparison with its competition, it must catch the eye. It must not be garish. It must sell the wine. This is a tall order. Also, the label must be approved in the USA by the Bureau of Alcohol, Tobacco and Firearms. There are several things that may be required for commercial printers. They must have four color process production, gold or silver embossing or printing. The die cut should be to a tolerance of 0.2 mm. The paper must be scuff proof, water resistant and tolerate high speed bottling lines. The price must be reasonable. The paper must be rough-sided for gluing, opaque so glue does not show through and pH-neutral. It is generally best to select a printer with extensive wine label experience.

Gushing

There are reasons for gushing when bottles are filled. These include microscopic mold growth. This can only be removed by drastic treatment. It occurs when bottles are stored outside or in damp locations in shrink-packs. Figure 7-11 shows a palletized shrink-pack of bottles. The mycelia form the focal points for bubble formation. Plastic sheeting inserted between the layer boards and mouth of the bottles will prevent infections. Another problem with shrink-packs is the etching of the glass by alkaline residues. Air trapped in these fissures will be a foci for bubble formation. An acid treatment of the bottles before storage will prevent this problem. Sometimes

FIGURE 7-11. Palletized shrink-packed bottles ready for the autounloader.

lubrication material or oil-like contaminants inadvertently get inside the bottle and cause problems.

The problem of gushing from opened bottles is not so clear and defined. The problem seems more related to red sparkling wines but also occurs in whites. Stress on glass to form enlarged micro fissures and some of the reasons above have been suggested. Luckily it happens rarely.

STORAGE AND TRANSPORT

Storage and transport of wines are not always under the winemaker's control. Considerable damage can be done to a bottled wine when it is subjected to excess heat or cold. The damage caused by cold is mainly breakage. When wine reaches its freezing point, around −3°C, it solidifies and then expands. The pressure can easily rupture the bottle and the wine is lost. If the bottle survives the freezing and doesn't rupture or push the cork out, then when the wine melts and all the compounds redissolve, the wine is essentially unchanged. If the wine was properly stabilized in the first place, it should be unaffected visually or tastewise.

However, heat can be disastrous to a wine. When the temperature gets above 40°C, it is only a matter of a few days before visual and sensory changes occur (Ough, 1985). SO_2 helps considerably in delaying the damages. After a time SO_2 can no longer protect the wine.

Common places where excess temperatures can occur are nonrefrigerated boxcars held on siding in the middle of summer or improperly ventilated warehouses where the heat accumulates near the roof. Other areas where heat damage happens are storage sheds (metal or other) where there is no insulation, cooling and ventilation or store windows in the full sun. This is not a complete list and any place where the temperature exceeds 25°C for long periods and over 40°C for short periods can affect wine quality. Another, more recent, aspect is that wines with urea present will more rapidly form ethyl carbamate. A complete analysis of the wine before shipment is a wise safety precaution in case damage occurs later.

HOME WINEMAKING BOTTLING

For home winemakers who will perhaps bottle their wine more simply, other precautions can be suggested. Clean bottles thoroughly with hot water and detergent if they are used bottles and rinse thoroughly. The wine should have been clarified satisfactorily by fining, racking and perhaps filtered and cold stabilized.

A rubber syphon hose that will comfortably fit into the neck of the bottle to be filled is necessary. The hose should be tasteless (soak overnight in a 20% ethanol solution to see if any off-tastes or odors are extracted). Food grade rubber is generally all right, but it is best to check. Start a syphon and fill the clean bottles. Capping the bottles with a crown cap is fine. They are as good as a cork for closing the bottle. If wine corks will be used, get the proper size. Be sure they are first quality and paraffined. It is necessary to have a relatively expensive corking machine to put the corks into the bottle. If the wine was dry and clean and contains about 20 mg/L free SO_2, it probably will be all right. Any precipitates that settle out can be removed by decanting the wine when used. Otherwise, everything said for the commercial operations would apply, especially the use of SO_2 and the care about high temperature.

Chapter 8

Sensory Evaluation

Sensory evaluation, more commonly called tasting, is the key to good winemaking. Knowing the ranges of color, taste and smell of wine allows the winemaker to determine the attributes and faults. Knowing the faults, one can correct the deficiencies or prevent further deterioration. Many times the changes occurring can be detected by tasting before they can be discerned by chemical analysis. Appreciating the good points of a wine is the whole reason for making wine.

EVALUATION OF THE WINE

The actual act of tasting a wine may appear ritualistic to the amateur, but each act has a significance.

Appearance

The first test is done by holding the wine up to the light. This serves two purposes. It determines if the wine is clear or hazy. Secondly, to observe the hue and depth of color of the wine (whether it is brown, green, yellow, red or purple and how intense). These can tell one much about the wine. If the wine is hazy, then there is reason to suspect either a microbial infection or some physical instability. If the wine is new and not yet clarified, then there is no real concern. If the wine had previously been clear, then there should be an effort made to determine the cause of the cloudiness.

The hue can also indicate undesired changes in the wine. Most white wine is light yellow or greenish-yellow at the start. If it darkens to deep yellow or brown, this is an indication of oxidation. This usually means aeration (or excess heat) has occurred. Red wines with purple shades of hue usually have a high pH. A red wine may either rapidly lose color or turn brown prematurely. This can indicate a microbiological problem or the wine has been subject to excessive heat or oxidation. Certain bacteria form dextrins. These make the wine "ropy." It strings when poured. Usually these wines have an opalescence (as do most heavily bacterially infected wines). When wines become gassy, it is an indication of microbial activity. If the wine is undergoing a desired malolactic fermentation, gassiness is expected. Otherwise it is a sign of spoilage.

Aroma

Once you are through with the visual examination, then it is time to smell the aroma. Here one looks for both good and bad notes. Some of the good smells you look for are varietal aroma, fruitiness and bouquet. Not all good wines necessarily have varietal aroma. The easiest detected varietal character associated with *vinifera* is that of muscat. This terpenol family of compounds is present in most grapes but in certain varieties they are much more concentrated. There are several different compounds which contribute to the odor (see Chapter 2). The mix of these are partially responsible for the difference in the varietal aroma of Muscat of Alexandria, Orange Muscat, Gewürztraminer and, to a lesser extent, White Riesling. The other main varietal aroma of *vinifera* is that of Cabernet Sauvignon, Sauvignon blanc, and Sémillon. This odor is due to compounds of the substituted pyrazine type (odor of vegetables such as green peppers and potatoes). These very odorous compounds are present in extremely small amounts. They can be detected rather easily at low concentrations. When present in too great amounts, they are disagreeable. Again, the mix of these compounds probably accounts for the differences between, say, Sémillon and Sauvignon blanc. Likewise, between the Cabernet varieties nuances of difference in character are probably because of differences in amounts of these compounds. One common red variety with a unique varietal aroma is Zinfandel. The raspberry character is very distinct when the grape is grown in cooler areas. Of the varieties which have distinctly different aromas are Chardonnay, Grenache and the Tinta Portuguese varieties. Other varieties do have distinctive aromas occasionally. Once one has these aromas firmly in mind, then one's ability to judge wines is greatly enhanced.

Other favorable aromas are the fruitiness derived from esters formed during fermentation. These esters, similar to those in all fruits, are produced by the yeast in abundant amounts. The esters disappear with time. The amounts of the esters that can be kept in the wine for a month to, at best, a year depends on the storage temperature. White wines retain the maximum ester retention and fruitiness if kept at 0°C.

Bottle bouquet comes about when aging changes occur. The rate of formation is increased by heat and reducing conditions (such as occur in wine without contact with oxygen). The complex compounds which are formed give a distinctive aroma. This aroma is sometimes described as a "burnt match" or "old cigar" odor. This happens in both red and white wines kept in closed containers. There are other distinctive changes which do occur. For example, Zinfandel looses it varietal aroma in time. White Riesling takes on a "kerosene-like" smell after a few years. This occurs especially if the wine is dry and picked past maturation. The aroma of wines made of grapes cleanly infected with *Botrytis cinerea* have a favorable note. In addition, the muted aroma of various oak flavors from barrels certainly is a good attribute. If a retsina is tasted, then the proper flavor of pine resin must be appreciated. To be a good wine taster you do not have to like a wine's style. You must appreciate and recognize good and bad attributes that relate to that style of wine.

The adverse aromas must also be searched for. There are many possible sources of off-aromas. A few are listed in Table 8-1 with their probable cause (others exist).

Taste

The next step is to put some of the wine into the mouth. In this step you further check the wine for volatile acidity, acidity, body, sweetness, bitterness, astringency, overall balance, and off-taste.

Table 8-1. Some Causes of Off Odor in Wines.

Smell	Cause
Vinegary	Acetic acid--ethyl acetate from acetobacter
Sewer gas	Hydrogen sulfide--reduced sulfur from yeast action
Oniony	Mercaptan--from H_2S reacting with ethanol
SO_2	Excessive use of sulfur dioxide
Sherry-like or aldehyde	Oxidation and/or excessive heat
Moldy	Presence of Penicillium mold on grapes or in a barrel
Rubbery	High pH grapes
Barn-like	Hot fermentation of red grapes, bacterial activity in the skins.
Corky (corked)	Mold activity on chlorine compounds used for preservatives of the cork material.

Each are thought about individually. This takes some amount of concentration. A good description of the process is available in Amerine and Roessler (1983).

The volatile acidity causes a prickly sensation in the back of the throat. To simulate this taste, take some vinegar and dilute it until you can get the sensation. On normal commercial vinegar, the label will state the percent of acetic acid present. Dilute it accordingly to about 0.1% for a first approximation. Then dilute further until you perceive the sensation without the very strong vinegar taste. Water solutions are less complex and odors and tastes are not masked to the same extent as in wine.

The acidity of a wine is normally attributed to the presence of tartaric and malic acids. To determine what the acid taste is, dissolve 0.2 grams of tartaric acid in about 100 ml of water and taste. The wine should have an acid taste that is sufficiently pronounced to be barely detectable but not so much as to be obviously tart. The perfect amount depends on the other compounds present which interact with the acid. These are mainly sugars, tannins and alcohol.

Body is caused in table wines mainly by the effect the ethanol has on the tongue. It is a warming (chemical) effect as well as a sense of the wine being heavier on the tongue. Excess sugar can also cause a heaviness in the mouth. Again, pure solutions of ethanol can be used to determine the chemical sensation. A solution of 8% v/v ethanol can be compared to a 14% v/v solution.

The effect of the sugars in wine are primarily that of sweetness. One hardly needs to make model solutions to demonstrate this taste. The threshold of sweetness detection in wines is around 1 g/100 ml as glucose. Some wines cover the sweet taste of sugar more than others. Levels less than these are detectable. The effect is not by sweet taste but by smoothing out the wine. The coarseness and other off-tastes are covered up. The acidity and tannin tastes are more balanced. You can easily demonstrate this effect by additions of sugar. Add 0.5-1.0 g/100 ml of sugar to wines of lesser quality with coarse tastes. The wine will usually taste better, but will not be sweet tasting.

Bitterness is usually thought to be the result of excess phenols. These come from ruptured grape seeds or stems extracted into the wine. Some bacteria can also cause the wine to be bitter. Again, this basic taste needs little demonstration. If one needs a substance for demonstration, then quinine is very bitter and can be obtained. The outer peel of the orange or lemon rind is also a bitter substance. If a wine is very bitter, then it is best discarded or distilled.

Astringency is the puckering feel or the roughness on the tongue caused by the more complex tannins reacting with the protein of the tongue. This drawing effect can be demonstrated by sucking on an unripe persimmon. The astringency of a wine comes partly from the wine and partly from the tannins extracted from the oak barrels the wine is aged in. There is a tendency for the astringency to decrease with time as the phenols become more complex and precipitate out.

The overall balance of a wine depends on many criteria. It is this general impression of this composition of integrated sensations that one has that is the flavor. The more harmonious the factors are in the wine, the better the overall balance. Nothing should be there in excess and nothing expected should be lacking.

Off-Flavors

Off-flavors can come from the grapes, fermentation, barrels or handling. Some of the flavors are given in Table 8-2 with possible sources.

There are other possible flavors, but these are usually detected as a smell and not as a taste sensation. Some of those in Table 8-2 may be questionable as a taste sensation. There are only four basic taste sensations and these are sweet, bitter, sour and salt. The latter is usually not a consideration in wine. Any other taste sensation is either a chemical effect or a combination of the four senses. It may be an actual smell that is being confused with taste.

SCORING SYSTEMS

There are numerous scoring systems used throughout the world. Perhaps the best known is the Davis Score Card developed at the Department of Viticulture and Enology at the University of California at Davis. It considers all the points discussed and can be used to

Table 8-2. Source of Some Wine Off-Flavors.[1]

Off-Flavor	Source
Hot or alcoholic	Overripe grapes--not overcropped
Flat, high pH, tea-like	Grape--overripe and overcropped, generally
Oxidized (rancid or insipid)	Aging--excess air contact or heat
Green or thin	Underripe grapes--insufficient ethanol
Metallic	Bacterial (or rarely excess metal pickup)
Bitter	Bacterial (tourne) or phenolic extracts from stems or seeds
Coarse	Grapes from region unsuited for variety

[1]Difficult to separate aroma from tastes; especially oxidized tastes

build up your skills as well as an excellent record describing the ever-changing wine. It is always good to include as many notes as possible as you taste the wine. If you taste the wine again in a month or a year, you will have forgotten the previous impressions unless they are on record. The Davis Score Card is shown in Figure 8-1. This is a slight modification of the version in Amerine and Roessler (1983).

Whether the Davis Score Card is used or another, keep good records and refer back to them after retasting a wine. This is espe-

FIGURE 8-1. Davis 20-point wine score card.

WINE SAMPLE NUMBER ________

CHARACTERISTIC	POINTS	(maximum)
APPEARANCE	______	(2)
COLOR	______	(2)
AROMA AND BOUQUET	______	(6)
TOTAL ACIDITY	______	(2)
SWEETNESS	______	(1)
BODY	______	(1)
FLAVOR	______	(2)
BITTERNESS	______	(1)
ASTRINGENCY	______	(1)
GENERAL QUALITY	______	(2)
TOTAL	______	(20)

Classification: superior (17-20); standard (13-16); below standard (1-12); and unacceptable or spoiled (1-8).

Tasters name ______________________________ Date ______________

cially valuable when tasting wines in barrels. The notes are sometimes more valuable than the number assigned to each component.

Coordination – Records

Coordination of sensory and chemical data is also essential. When you suspect a change in the wine, do an appropriate chemical or microbiological test to verify your taste judgement. This builds up your confidence if you are correct. Even more importantly, it teaches you what aromas and flavors are associated with the various chemical, biochemical and microbial changes. It is most gratifying after a few years to be able to taste a wine and *know* what is wrong (or right) about the wine and why.

Proficiency

It is perhaps easy to see why anyone hoping to be a proficient wine taster needs at least five years of continual tasting experience. One has to build up a knowledge of all the possible sensory sensations of the wines. Also, one needs to catalog them in his/her mind. One, when evaluating the wine in question, must pull out matches from previous tastings to make decisions.

Wine quality judgements are subjective. For the person who doesn't taste wines and evaluate them continually, it may appear a rather unscientific exercise. However, tasting of wines for quality by experienced wine tasters can be rather exacting.

Wine scoring systems have no magic. Any scoring system is adequate if the user is familiar with it. The experienced taster will use the scoring sheet mainly to help him/her concentrate on the various aspects of the wine. He/she will assign a value number that represents his/her overall evaluation of the wine. This may not necessarily be the sum of the individual points that were assigned to certain attributes.

Any starting enologist who wants to become an expert in his or her chosen field should become familiar with Emile Peynaud's (1987) book *The Taste of Wine*. This comprehensive text relates the experience of one of the world's foremost enologists and puts the sensory evaluation of wines into a true perspective. One passage is memorable and worth quoting for beginners:

> I [Peynaud] often tasted with [Jean] Ribéreau-Gayon, at a certain period practically every day, and first of all on a tasting bench in the laboratory. We forced ourselves to practice tasting even though to begin with, it seemed obscure, uncertain and difficult to reproduce results. Progress remained slow until one had acquired a reliable technique and a basis for comparison. Even in such favorable circumstances it took us years to synchronize our tastes and acquire the necessary reflex reactions to agree about the meaning of words.

Nothing worthwhile comes easily. There are no shortcuts or perfect words, easy statistics, etc., only long, hard work to become a proficient wine taster.

DESCRIPTIVE TERMINOLOGY AND ANALYSIS

Terminology in the wine business is not uniform. Evidence of the need for standard terminology to describe wine aroma and flavors was the plea of Razungles and Bidan (1987) and the suggested score card of Brigaud and Medina (1987) which tries to include some descriptor terms and ratings. This will certainly help in describing the wine and standardizing the descriptors.

Jordon and Croser (1984) give details on assessment of potential wine flavor by sensory evaluation of the juice. They point out that oxidation of the must be minimized to prevent the formation of the C_6 aldehydes which can mask the fruit aromas. To do this, SO_2, ascorbic acid and pectic enzymes are put into the receiver and the berries crushed directly into the receiver. The juice is cooled to near 0°C, sparged with nitrogen, SO_2 adjusted, sealed in a bottle prefilled with CO_2 and cold settled for 24 hrs. Then decant the clear juice and carry out the desired aroma, flavor and analytical tests. The samples can be saved for several weeks. This testing is done for several weeks before anticipated harvest date. As the grapes approach optimum flavor and aroma maturity, the decisions to pick can be made based on the characteristics desired. Some of the terms used by the authors are noted in Table 8-3.

The assessment is carried out carefully and rigorously with statistical validation of the intensity differences noted between maturity

Table 8-3. Changes in Aromas Associated with Various Cultivars.

Maturity	White Riesling	Sauvignon blanc	Chardonnay	Cabernet Sauvignon	Shiraz
Immature	Green/ unripe	Green/ unripe	Green/ unripe	Green/ unripe	Green/ unripe
Partially ripe	Green/floral	Light/herbaceous, grassy	Cucumber	Light herbaceous	Herbaceous, spicy
Fully mature	Citrus-lime, perfume-floral, herbaceous tropical fruit muscat	Strong herbaceous, capsicum, tropical fruit	Cashew, tobacco, melon, ripe fig	Grassy, herbaceous, capsicum, dusty/earthy, black currant, blackberry	Spicy, peppery, confectionary,raspberry, jammy

samples. It can be invaluable in not only optimizing picking times, but for evaluation of vineyards. A vineyard that doesn't produce flavorful grapes will not produce flavorful wines. Comparisons of juices between vineyards offer considerable additional information in deciding what fruits will be purchased in the future.

Descriptive analysis, for example, can statistically show a wine has certain attributes. It can define the aroma and taste in spider diagrams very well. How well all the attributes blend to make a wine of a certain quality it cannot do. An intelligent person should use all the tools at their disposal to evaluate a wine. Descriptive analysis lends itself well to describing the aroma. From this, a great deal can be discerned. Wines from vineyards of the same cultivar grown only short distances apart can be shown to be distinctively and statistically different. This can be done by either a descriptive analysis or a quality evaluation scoring. The description tells something of why they are different and the quality evaluation says which is better (as judged by experienced tasters).

Descriptive analysis does not require a great amount of experience. Young, sensitive subjects are probably better qualified because of their sensory acuity. Brief training sessions can familiarize them with the few various odors and tastes associated with a specific wine type. The descriptive terms which the operator wishes them to use are limited. The aroma wheel (Figure 8-2) for wine prepared by Nobel et al. (1987) uses terms that are realistic and descriptive to define wine aromas. In addition, they provide a method to produce most of the aromas. This is a needed effort to bring uniformity to terms used to describe wine aromas. Words

FIGURE 8-2. Modified ASEV Wine Aroma Wheel showing first-, second-, and third-tier terms.

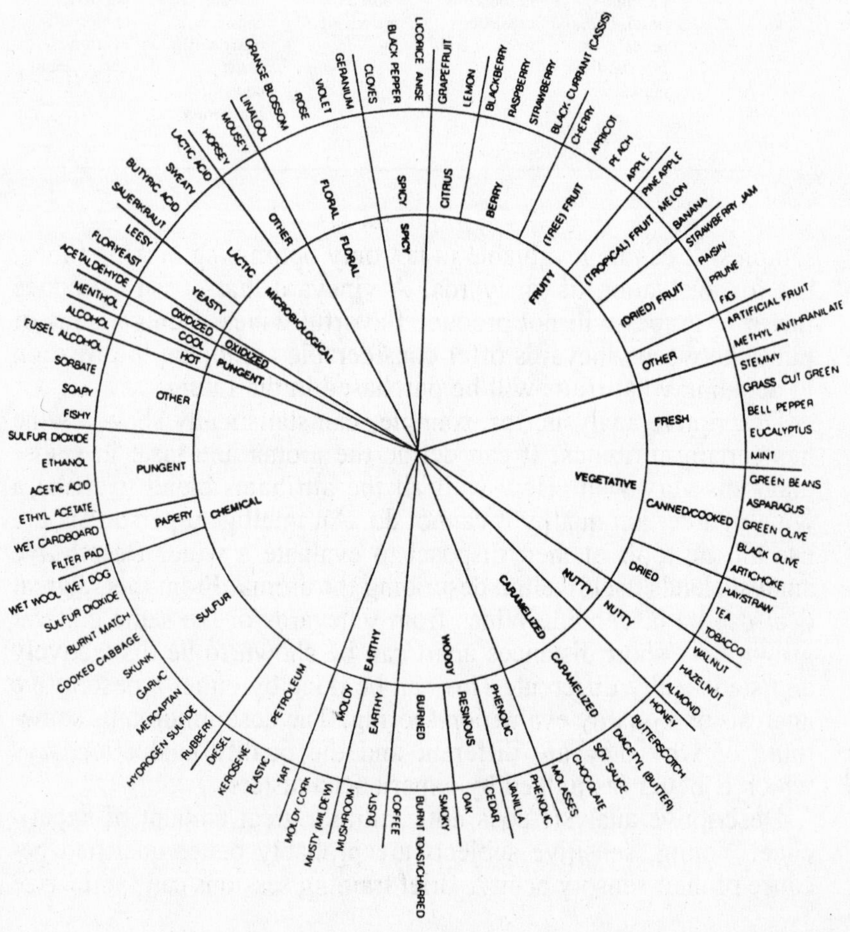

which do not describe the wine should not be used. Such terms are subtle, gay little wine, moody, upright, forthright, charming, etc.

This type of analysis does not allow you to determine the quality value of a wine. It just describes certain attributes but does not give you the overall integrated impression needed to make certain critical decisions.

STATISTICS

Much emphasis is placed on statistical evaluation of results. This is essential to determine if data means are different or just random dispersion. It is foolish to think that any set of data gathered from human subjects is objective or less subjective than another set of data. There are two criteria that can be considered: (1) Is the answer a yes or no answer? (2) Is the answer one of scoring, scaling or rating? The yes/no decisions are by far the easiest to make. One either detects a substance or one does not. Simple Student-T statistics can be used to determine statistical significance. Similarly with paired samples, duo-trio tests or triangular tests, the questions can be made specific for a criteria and the analysis straightforward.

In judging wine for absolute criteria or quality, the statistics are less simple, though similar. Analysis of variance is used for scientific work. Scores or ratings made independently are replicated. The variability of tastes and interactions are separated and the treatment variance determined and compared to the proper error-variance. Differences in treatment means can be calculated. For those wishing a more detailed treatment of wine evaluation, see Amerine and Roessler (1983).

Descriptive analysis is treatment of the terms used about the treatment variable(s). The variable components of significance are plotted to show separation and importance. An example of such a spider or cobweb diagram is given in Figure 8-3 from Goniak and Noble (1987).

These data can be further treated by principal component analysis. Factors are determined that contribute to the aroma quality for each wine. In this sense, quality refers to an identifiable component(s) intensity not to final or overall wine quality.

FIGURE 8-3. A "Spider" or "cobweb" polar coordinate graph of mean intensity ratings and least significant differences (LSD) of dimethyl sulfide (___) and ethanethiol (....) for the lowest (L) and highest (H) concentrations (From Goniak and Noble, 1987).

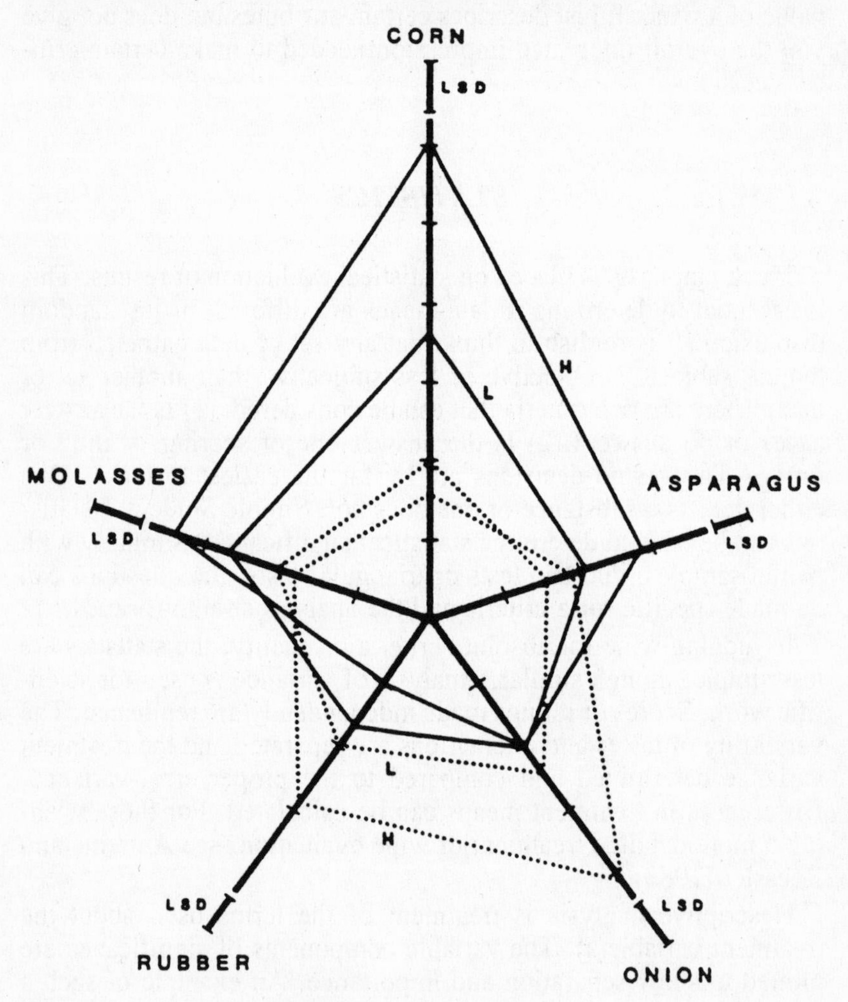

WINERY TASTING

Winemakers, especially in smaller wineries, have very little opportunity to do the more elaborate statistical procedures. They do not have the time and people to do the extensive number of replications necessary to prove the variable has a significant effect. Instead, most wineries depend on one or a few highly experienced tasters to make many decisions. Among these are: blending judgements, the determination of when wines in the barrels are to be removed, the decision as to the quality value of a wine they are about to buy or sell, compare their wines to the competitors' wines, check wines in barrels and tanks for spoilage and evaluate wine made from different vineyards and areas. All these jobs are done usually in a simple tasting situation. The economic life or death of the winery depends more heavily on these efforts than almost any other winery activity. Good winemakers are conscious of this. They make every effort to taste at every opportunity, keep notes of tastings and to join to taste interesting wines that the group may have.

Tasting only your own wines is a folly. You may get used to some spoilage characteristics that has come from infected tanks or barrels. It may just become part of what you term wine aroma, but it could be what most others would classify as spoilage. It is extremely wise to compare your wines carefully with those of your competitors and occasionally get an outside unbiased opinion of your wines. There is always something to learn from listening to others describe wines. Winemaking styles change. Each impression and opinion will help formulate what one does in the future as a winemaker as well as how one evaluates wines.

Chapter 9

Chemical Analysis and Information Retrieval

ANALYSIS

The need for extensive analyses of wines made for home use is seldom necessary. However, if a better appreciation of the wine is desired, then the grape and wine analyses are extremely helpful. When problems arise or a particularly good wine is made, analyses should be made. These may partially explain the failure or success, and help in improving future wines. There is no legal requirement for any analyses for wines produced for home use. For commercial wineries, these analyses are the basic minimum and most competent winemakers will either do or have done these and other analyses as needed.

There are a few basic analyses which can be done easily with limited equipment and expense. These are °Brix, total acidity, pH, ethanol, volatile acidity and SO_2 (free and total). Iron, copper and total phenol analyses can be done but do require some special reagents or equipment. Unless the winemaker has some experience in this area, these are best sent out to a commercial laboratory.

A list of common units in metric and American conversions are given in Appendix II.

GRAPE JUICE

The chemical estimate of the sugar acid, and pH of the grapes are fairly simple. The analysis can be no more accurate than the sample taken. It is therefore wise to follow the procedure outlined in Chapter 1 or any version that gives consistent and accurate results for grape sampling. Once you have the proper sample, then you proceed with the analysis.

°Brix

A refractometer is a simple tool that can be used to estimate the sugar content of the grape by measuring the refractive index of the juice. The refractometer (Figure 9-1) is calibrated as g sucrose/100 g of solution. Usually the °Brix reading of juice is approximately 1.5-2.0 g/100 g more than the actual sugar content. The sample of juice is placed on the prism face of the refractometer and the cover closed. Be careful not to use a hard applicator that can scratch the glass prism. The refractometer is held up to the light and the field observed. There is a scale inside which usually is from 0 to 30°Brix. The field will be divided into a dark and a light field. The junction of the two fields will cross the scale at the °Brix of the juice. Most of the modern refractometers are internally corrected for temperature effects. Excessive variation from the temperature of 20°C should be avoided as the compensation may not be accurate much beyond ± 10° from 20°C in some instruments. Temperature corrections for the refractometer are given in Appendix III.

FIGURE 9-1. Handheld refractometer for field measurements.

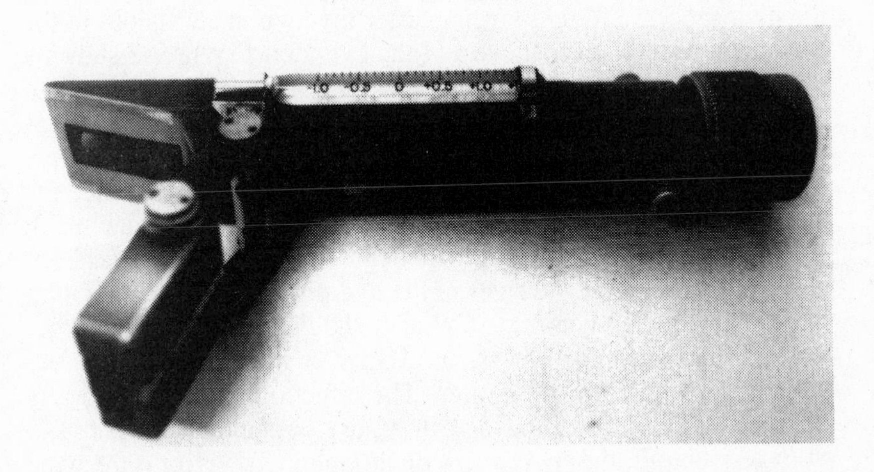

Hydrometers are also calibrated as °Brix or g sucrose/100 g solution. These are available with various scales. The best kind has a range of approximately 10°Brix. Three or four hydrometers cover the range needed for measuring juice as well as following the fermentation. (The refractometer *cannot* be used to follow the fermentation because of the refractive index of the ethanol produced.) Have the juice reasonably clear before using the hydrometer. This can usually be accomplished by a few hours of settling and then racking off the cleared upper layer or centrifuging. There is a temperature correction needed if the temperature of the juice varies from 20°C (the usual calibration temperature of the hydrometer). This correction can be approximated by adding 0.1°Brix for every 1°C above 20°C and subtracting the calculated amount if the temperature is below 20°C. For a more exact correction, see Appendix IV. The correction is no more accurate than the temperature reading.

If the fermentation is being followed, then there will be some interference from the carbon dioxide gas escaping. This gas can cause the hydrometer to sink further than is justified by the actual °Brix of the solution. If the hydrometer is left in the fermenting wine too long, the bubbles will adhere to the surface of the hydrometer and cause it to ride up higher than the true reading. Good results are obtained by putting the fermenting juice into the cylinder. It should be sufficiently large in diameter so as not to restrict the hydrometer. Fill to a level so after the hydrometer is put in the liquid, the level is near the top of the cylinder. Figure 9-2 shows a plastic overflow hydrometer cylinder. The hydrometer is put into the fermenting juice and spun gently several times. This will degas the sample enough to allow the hydrometer to come to a reasonable equilibrium. Make the reading within 10 or 20 seconds after the hydrometer has reached a stationary level. (If one waits much longer, the hydrometer will slowly begin to rise due to the gas bubbles accumulating on the sides of the hydrometer.) The temperature of the wine in the hydrometer cylinder should be noted at that time. As the wine reaches dryness, the °Brix reading will be negative. Ethanol is less dense than water. The effect of the alcohol lowers the specific gravity of the wine below zero. When the wine is dry or stops fermenting, the °Brix will stop dropping. To verify if the wine

FIGURE 9-2. A plastic hydrometer cylinder with an overflow basin.

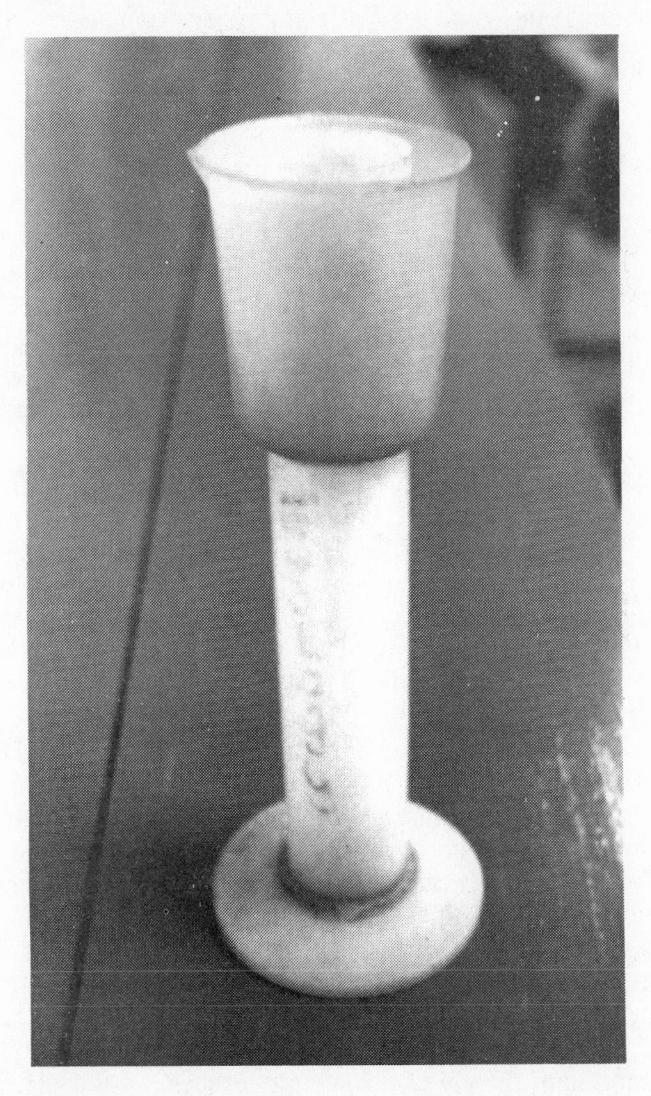

is dry, a Dextro Check® tablet can be used to determine the sugar level. These are tablets produced by the Ames Company. They will estimate the reducing sugar accurately between 1 g/100 ml to 0.1 g/100 ml. When the reducing sugar (fermentable sugar) gets to 0.2 g/100 ml or less, the wine is essentially dry and ready to rack. There are other hydrometer scales used in various parts of the world but the principle is the same–specific gravity determination. Tables or calculations can be used to convert one scale value to another.

pH

The small portable pH meters are not too expensive and are a good investment if one is serious about winemaking. Standardize the meter using a solution of potassium hydrogen tartrate (potassium bitartrate). Put a small amount (sufficient to saturate the water) of this material into a beaker with about 50 ml of water. The pH of this solution will be 3.56 at 20°C temperature. Most grape or wine measurements will be within half a pH unit of this value.

Total Acidity

The total acidity or titratable acidity is measured by titrating a measured degassed sample to a pH end point of 8.4-8.6 with standardized sodium hydroxide. The end point can be determined with an indicator, phenolphthalein, or (easier) with a pH meter. The wine or juice can be degassed of CO_2 by either putting it under a vacuum and shaking or adding 100 ml of boiling water to a wine aliquot or by bringing the aliquot of wine and the water to an insipiant boil. In the latter two instances, if the end point is to be determined by a pH meter, the solutions must be cooled first. Standardize the pH meter with a buffer of about pH 8.0 value. Buffers are available from wine equipment shops or from laboratory supply houses. Keep the buffer in the refrigerator. Inspect it for mold or other microbial growth which will alter the pH of the buffer, making it useless. Add about 100 ml of boiling water to a 10 ml (exactly) portion of the juice or wine in a 250 ml beaker. Place the electrodes of the pH meter in the cooled solution. Place a magnetic stirring bar into the solution. Put the beaker on a magnetic stirrer and stir at a medium speed. Titrate the solution with standardized (0.05 *N* NaOH) from a 25 ml buret until the pH of the sample reaches 8.4-8.6 pH. Figure 9-3 shows a typical setup. Record the amount of sodium hydroxide used. Use the following calculation:

$$\text{g tartaric acid/100 ml} = \frac{\text{ml NaOH} \times N\,\text{NaOH} \times 75 \times 100}{\text{ml wine used}}$$

FIGURE 9-3. A setup for total titratable acidity determination using a pH meter to detect the end point.

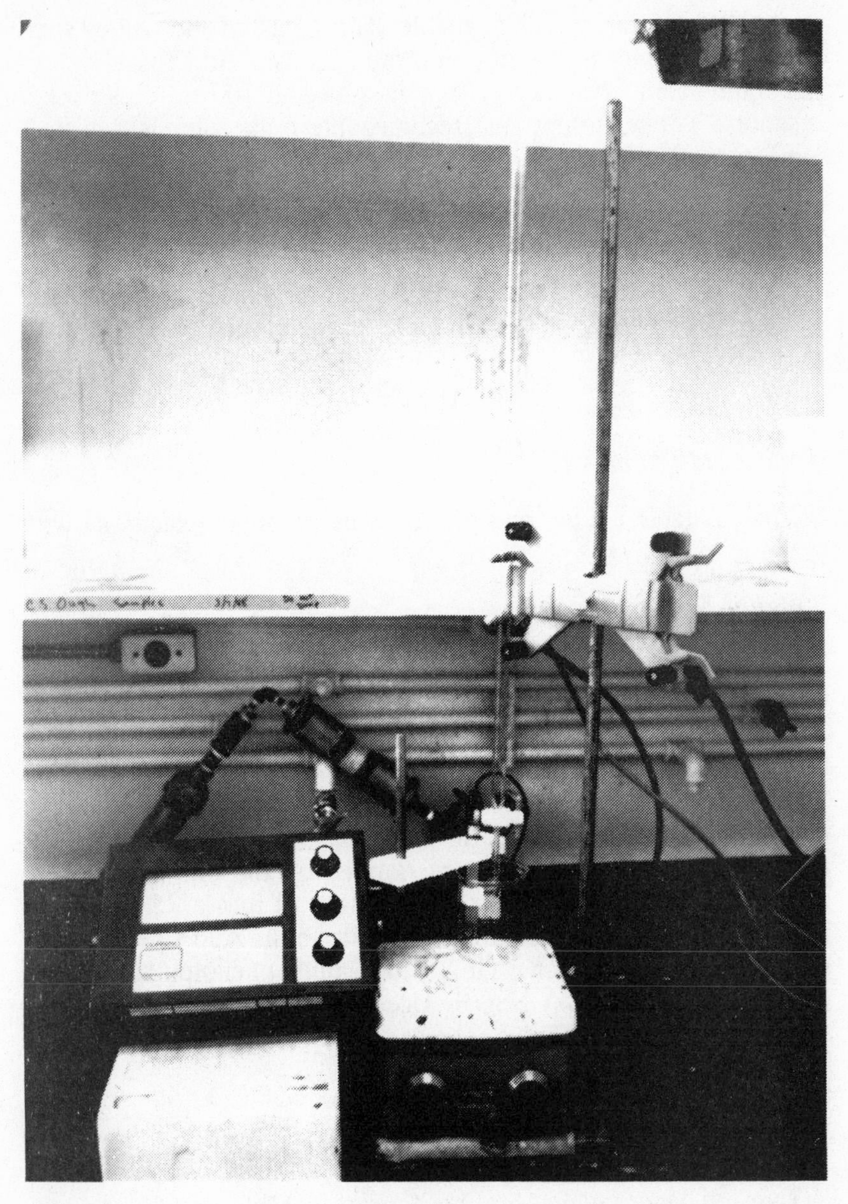

Be sure that the water you use to dilute the sample is neutral. If it is not, neutralize to pH 8.5 before boiling for use.

A pH meter may not be available. The degassed and diluted sample is placed into a 250 ml wide mouth flash. Add five drops of phenolphthalein indicator (1.0 g dissolved in 100 ml of 95% v/v ethanol). Titrate until a just recognizable pink color is detected. With red wine it takes some experience to detect when the color change occurs. Pigments can mask the color changes. The two critical measurements are the exact wine volume and the exact milliliters of NaOH. The calculations are the same as above. Automatic titrators are available and useful for large wineries where many samples are to be done. An example of one is seen in Figure 9-4.

WINE

Total Acidity and pH

The titratable acidity and pH for wine are done exactly as with grape juice.

Volatile Acidity

The volatile acidity requires a glass still. Figure 9-5 gives an example of the Cash still. The Cash still is partially automated. It is necessary for good work. Fill the bottom of the still with distilled water to the side mark. Pipe 10 ml of wine into the inner chamber. Boil the water in the bottom chamber. The volatile acids are steam distilled over and condensed in the water-cooled condenser. Most of the other distillable acids are trapped by the one plate reflux above and do not distill over. Distill 100 ml into a 250 ml wide mouth flask. Bring the distillate to a near boil. Add five drops of phenolphthalein solution. Titrate to a pink phenolphthalein end point with about 0.02 *N* NaOH. Use a 25 ml buret. The end point should be a faint pink color. The calculation is as follows:

$$\text{g of acetic acid/100ml} = \frac{\text{ml NaOH} \times N\,\text{NaOH} \times 60 \times 100}{\text{ml of wine used}}$$

Two substances which can interfere, if present, are SO_2 and sorbic acid. Normal amounts of SO_2 can be ignored for practical pur-

FIGURE 9-4. Automated total titratable acidity and pH determination.

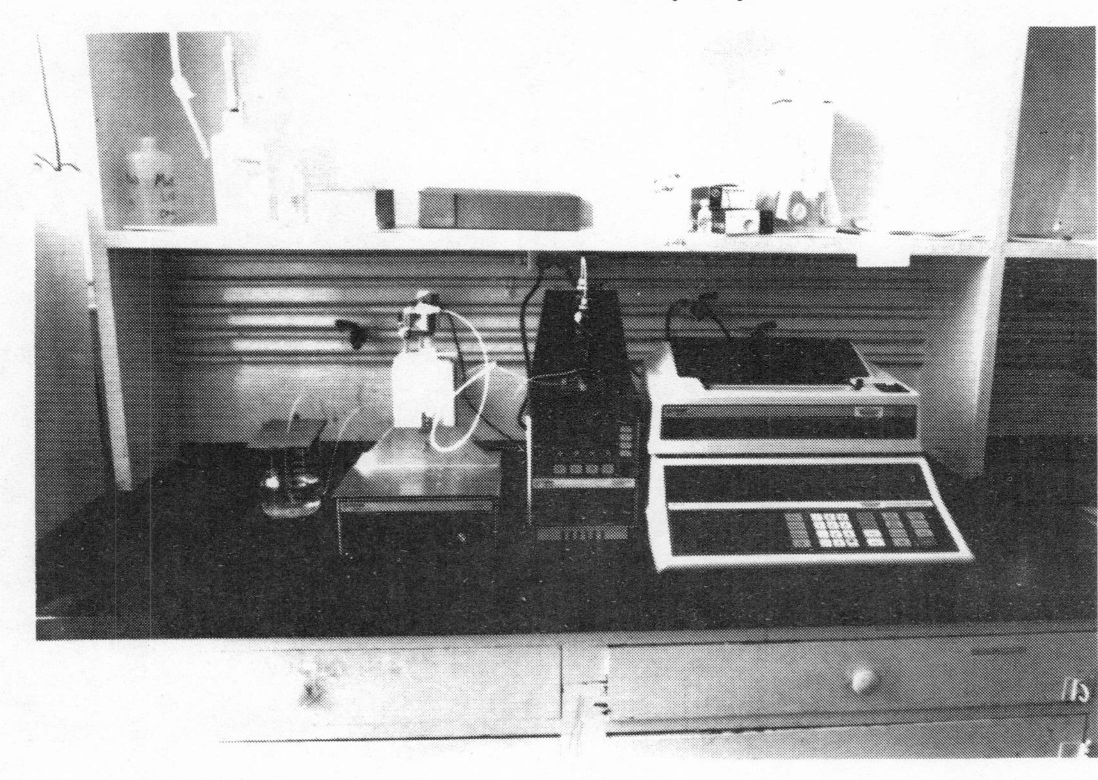

FIGURE 9-5. A cash volatile acidity distillation apparatus.

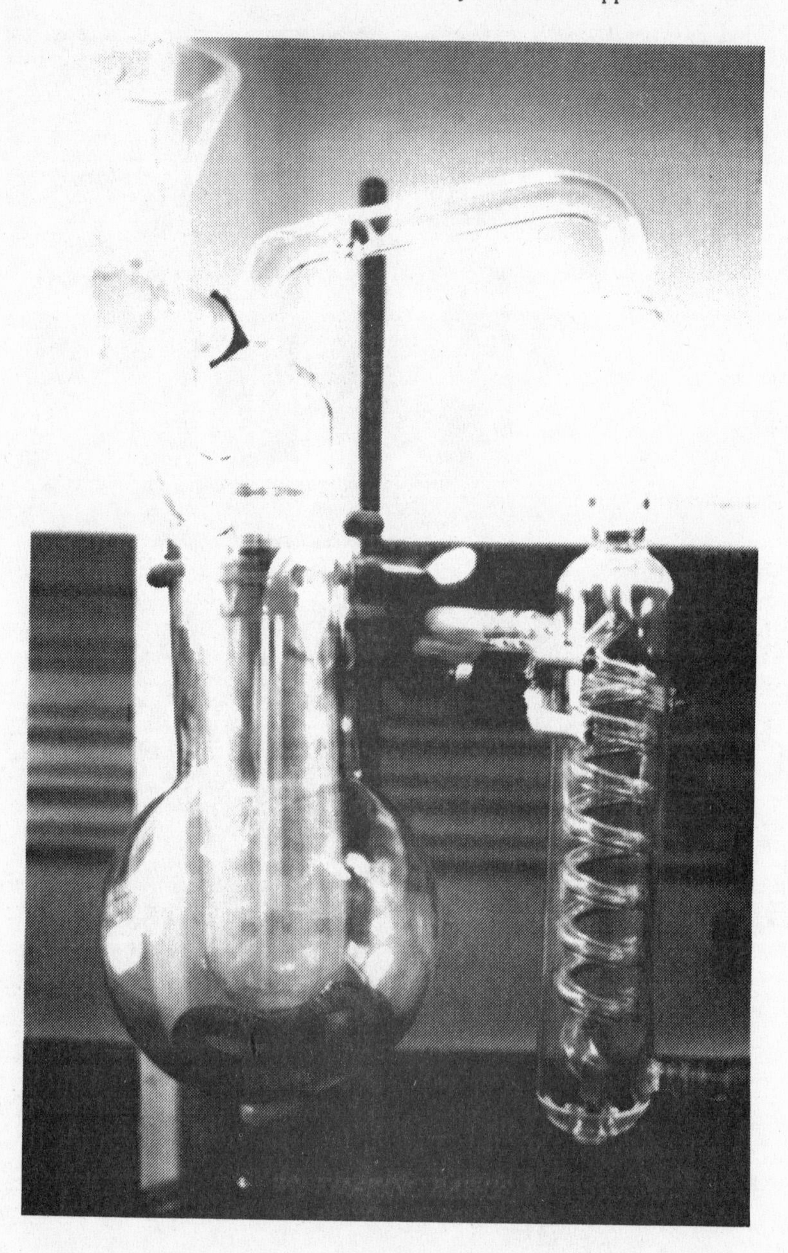

poses. Do a sorbic acid analysis if sorbate has been added to the wine. Calculate the amount as acetic acid and subtract from the value. However, few wines approach the legal limits and cause the corrections to be used (Ough and Amerine, 1988).

Ethanol

Ethanol or ethyl alcohol can simply be determined by the use of an ebulliometer. This is a standard device used to determine ethanol concentration in wine by measuring the boiling point. The boiling point depends primarily on the percentage of ethanol present.

These devices are available at wine supply shops and some laboratory supply stores. It is best to get a BATF-approved model. Some cheap glass versions are available but they are extremely erratic and give inaccurate results. The proper metal ones will cost more than $200. With proper care, these will last a lifetime and give accurate ±0.2% and reproducible results for dry wines. Directions usually come with the ebulliometer as each type has a slightly different volume consideration. The most common type is the Dujardin (Figure 9-6). The method for using this type is as follows.

Add 15 ml of water to the bottom part. Fill the condenser with cool water. Place the lighted alcohol burner under the hood on the drain spout. Place the thermometer into the head space of the bottom part. The water will boil. The thermometer will reach an equilibrium temperature. Set the calculator on that temperature value for 0.0% ethanol. It can vary slightly depending on the barometric pressure. Drain the water. Rinse the bottom portion with 15 ml of wine. Drain and then add 50 ml of wine. Refill the condenser and place the thermometer back into its place. Heat to boiling. When the temperature is stabilized, take the reading. (Do not wait too long as, after a time, the condenser water will get hot. The ethanol vapors will escape, causing the temperature to rise gradually.) Refer the temperature of the boiling point of the wine to the calculator. Read off the ethanol value as % v/v. There are other good methods; some more accurate. For these, refer to Ough and Amerine (1988).

FIGURE 9-6. A Dujardin ebulliometer for rapid alcohol determinations.

SO_2

Sulfur dioxide measurements can also be done by several different methods. However, the simple Ripper determinations are easiest and fairly inexpensive. The glassware is minimal and the solutions needed can be easily purchased or made if one has the knowledge.

Keep stock 0.2 *N* (or about) iodine solutions in brown bottles tightly closed until ready to use. Dilute the stock iodine solution tenfold. Dilute sufficient for a day's use. Standardize that with thiosulfate solution. This is purchased standardized at about 0.05 *N*. Take 20 ml of the approximately 0.02 *N* iodine solution. Add 5 ml of 9 *N* sulfuric acid, 5 ml of starch solution. Titrate to a very faint blue end point. It will be dark blue to start and, as the iodine disappears, will gradually become fainter. If you go past the end point, it will be colorless. The calculation for the standardization is:

$$N \text{ of iodine} = \frac{N \text{ thiosulfate} \times \text{ml thiosulfate}}{\text{ml of iodine solution}}$$

Determine free SO_2 by pipetting 25 ml of wine into a 250 ml Erlenmeyer flask. Add 5 ml of a starch solution and 5 ml of 9 *N* H_2SO_4. Titrate with 0.02 *N* iodine solution to a faint blue end point. Use a 25 ml buret. The end point of the juice or wine titration will fade. An accepted criteria is if the end point faint blue lasts for one minute. The free SO_2 calculation is:

$$SO_2 \text{ mg/L} = \frac{\text{ml iodine} \times N \text{ iodine} \times 32 \times 1000}{\text{ml of wine}}$$

Measure total SO_2 by first adding 5 ml of 12 *N* NaOH to 50 ml of wine in a 250 ml wide mouth Erlenmeyer flask. Close the flask with a rubber stopper for 15 or 20 min. Add 10 ml of starch solution and 10 ml of 9 *N* sulfuric acid. Titrate to a faint blue end point with standardized (approximately 0.02 *N*) iodine. The sodium hydroxide is added to release the bound SO_2. Calculate the same as for the free, with attention paid to the different amount of wine used. The aeration-oxidation method is superior, but more time consuming. For discussion of this and other methods, see Ough and Amerine (1988).

WATER TESTS

Astley and Rankine (1985) give simple tests for water hardness and available chlorine content.

Hardness Test

Put 100 ml water into 250 ml Erlenmeyer flask. Add 2 ml 0.1M NH_4OH buffer and a few grains of solochrome black indicator. Titrate with 0.02 *N* EDTA from a reddish pink color to the blue-gray end point. Hardness is as mg/L and is obtained by multiplying ml EDTA used by 10.

Chlorine Test

Pipet 25 ml of water into 250 ml Erlenmeyer flask. Bring the solution to volume with distilled water. Pipet 10 ml of this solution into a 250 ml Erlenmeyer flask. Add 10 ml of 10% KI solution, mix and add 20 ml of 1:9 sulfuric acid. Titrate immediately with standardized 0.100 *N* sodium thiosulfate until the yellow color of iodine is just perceived. The amount of available chlorine is given in g/100 ml by multiplying the ml of thiosulfate used by 0.36. To get mg/L, multiply answer by 10^4. For powders, weigh out a suitable amount and dissolve in 50 ml water in an Erlenmeyer flask. Proceed as above. Divide the answer obtained by half the weight of the powder used to get other analyses wt/vol value (use proper units).

Some home winemakers will not go to the trouble and expense to do these analyses. Others will and a few will want to get more involved. For those, the text mentioned previously is a good reference source for analytical procedure concerning juice and wine.

Any winemaker in a commercial winery will need far more expertise in chemistry than is needed for these simple analyses. However, these discussed include some of the ones that must be correct and be within certain legal tolerances (volatile acidity, SO_2 and ethanol). Other analyses commonly done are sorbic acid, iron and copper in table wine wineries. Total phenols are of interest and more and more wineries are investigating the various phenols found in wines.

Corrections for SO_2 must be made using the Folin-Ciocalteau method for phenols. According to D'Agostino (1986), it is the only workable method available if the must has been heated. This method is available in several texts including the one mentioned earlier.

INFORMATION RETRIEVAL

In the past, if a winemaker wanted to know something about wine technology or regulations, he/she had few options. He/she could consult a text on the subject, review his/her meager collection of journals or reprints, call someone who might know or drive to a major university library. Considering these options, most winemakers read the text and perhaps consulted with the other winemakers in the area when dealing with a problem.

With the advent of computers, this should no longer be the case. Any winery can afford a computer. The ease and information available for record-keeping alone more than pays for the few-thousand-dollar investment. Beyond that, the latest scientific information can be at one's fingertips. With a modem, the computer can be telephone-connected to sources of information. These sources are up to date and complete for enology and viticulture research publications on all subjects. There are two major sources. Scientific and Technical Information Network (STN) contains all the chemical abstracts (American Chemical Society). These are referenced from 1967 on. They have over 60 other data bases. The other source is DIALOG. This has the AGRICOLA, Bibliography of Agriculture (Dept. of Agriculture), Food Science and Technology Abstracts and Biosis Previews Abstracts. These two services cover most of the publications worldwide on grapes and wine. Vitis-VEA is included in Food Science and Technology abstracts. It deals solely in viticulture and enology abstracts. DIALOG gives abbreviated abstracts with citations in English and language of origin. DIALOG is based in Santa Clara, California. STN service is based in Columbus, Ohio. The use of the systems is simple but must be used with care. A search should be well planned and thought out. For example, with STN and using CAS ONLINE, the number of citations is given to the researcher before any printing is done. If there are too many, the search can be tightened up. For example, if someone was interested in SO_2 analysis methods they might start by the sequence:

SO_2, sulfur dioxide, analysis, methods, last three years.

If 526 citations came up, this would be an expensive and wasteful printing and difficult to go through. So the word "food" could be added to the criteria for search. This might cut it down to 150. Depending on the need, the words wine and grapes could be added and then the number might be reduced to 22. This would give you the abstracts of all the papers dealing in SO_2 methods of analysis for grapes and wine for the last three years. Off-line printing of searches are cheaper than direct responses. STN has the complete chemical abstracts while DIALOG has only the citation and a few key words for the abstracts. If any new methods had appeared and were of interest, then the papers could be obtained from either the authors or a library for further investigation. The same technique could be applied for almost any aspect of winemaking and grape growing.

Chapter 10

Additives and Contaminants

The purpose of this chapter is to clarify why certain additives are useful to wine. Risks from the use of illegal ones are discussed. Other additives used in the past and not in the present are reviewed. Some used elsewhere, but not approved for use in the USA, will be covered. Contaminants will be discussed.

ADDITIVES IN GENERAL

Additives are those components which are purposely added to a food product for a specific purpose. Contaminants are those unwanted components that either remain from some previous exposure or accidentally get into the product.

Farrer (1987) lists general additive types used in foods. Table 10-1 is a slight expansion of his list. Acids, aeration, antioxidants, color, enzymes, minerals, preservatives, sweeteners, clarifiers and gases are on the list. They are all legal to use in grape juices or wines. The acids and sweeteners are those normally present in the wine. They come from the grape. The clarifiers are usually removed after use. The preservatives are limited to just a few in use: SO_2, sorbic acid and dimethyl dicarbonate. Enzyme general use is limited to pectic enzyme treatment of the grapes, although others may become accepted.

Carbon dioxide and nitrogen gases are permitted. The only antioxidants used are sulfur dioxide, ascorbic acid or erythorbic acid. Air can be used. Copper is added to remove hydrogen sulfide. Caramel is the only legal coloring agent. It is seldom used in table wines. Recently the BATF published (Fed. Reg., 1990) the revised list of approved additives for wine.

Table 10-1. Some General Food and Beverage Additives and Their Purposes.

Additive	Appearance	Texture	Keeping quality	Flavor	Nutritional value
Acid, base or buffer			+	+	
Aeration	+	+			
Anti-caking, flow properties	+	+			
Antioxidant			+	+	
Color	+				
Emulsifiers	+	+			
Stabilizers	+	+			
Enzymes		+		+	
Flavors				+	
Humectants		+	+		
Minerals					+
Preservatives			+		
Sequestrants			+		
Sweetness				+	
Vitamins					+
Clarifiers	+		+		
Gasses	+		+	+	

Contaminants

What happens to the grapes before they reach the winery is not always under the control of the winery. Also, accidents can occur which can contaminate the juice or wine at the winery. The winery personnel must be aware of potential problems and minimize risks.

Pesticides, fungicides, etc., may be on or in the grapes at levels higher than regulations permit. It is the winery's responsibility to analyze the wine (or grapes). They must be sure that the levels are below the accepted legal standards. The farmer may be ultimately libel for losses. However, the bad publicity associated with the detection of pesticides in a company's wine is costly. Damages are unrecoverable in most instances.

Solvents, or similar materials, from improperly prepared tanks or tank linings can contaminate a wine and leave unacceptable residues. Oil or hydraulic fluid spills in gondolas or from equipment cause unwanted odors and tastes. The use of non-resistant metals such as aluminum, copper, etc., can cause high levels of these metals to build up. Breaks in chiller tanks or lines that can allow ammonia or glycol into the wine must be detected early. Fix the potential leaks before contamination occurs.

No food product is completely free of insect parts and similar debris from the field. Proper care to minimize this type of unwanted material should be carefully exercised. Detection of any part of an insect in a tank of wine can cause the wine to be seized. The California Public Health Department can destroy the whole tank as contaminated. This is a very severe penalty. A small amount of care and money can be well spent in avoiding problems of this nature.

LEGAL ADDITIVES

Sulfur Dioxide

Sulfur dioxide is probably one of the oldest antiseptics used. The death of insects that flew through SO_2 vapors occurring around active volcanos was observed centuries ago. This observation probably caused the first inspiration to burn a sulfur wick to sterilize a vermin-infected domicile. It is on the written record that the Greeks were familiar with SO_2 properties in this respect.

In use in wine, the purpose is mainly twofold: (1) to inhibit yeast and bacteria, and (2) to prevent oxidation. It is excellent in both respects. These effects were discussed in some detail in the earlier chapters.

The medical effects of SO_2 must be considered and understood. SO_2 is a very strong bronchial constrictor. Even a person with excellent health can find the sensation of breathing SO_2 fumes frightening. To a person with asthma or other lung dysfunctions, it can be deadly. A question that should always be asked of new winery employee is: Do you have asthma? If the answer is yes, then he should be warned of the danger. Not only do asthmatics have problems breathing but they can go into anaflaxtic shock. Without proper treatment, they can die. The breathing of the vapors of SO_2 is more serious than drinking the same amount in a beverage. The reason for this is that the SO_2 in air is all in the SO_2 state and is absorbed directly into the bloodstream. When it is ingested orally, the degree of ionization depends on the pH and the bound SO_2 will remain bound in the stomach.

The FDA considered whether to specify on a bottle of wine that the product contained SO_2. They directed that a study be done to determine if bound SO_2 contributed to the problem with asthmatics. It was found by Simon et al. (1988) that a portion of the subjects tested did respond to bound SO_2. This response was about 10-15 min later than to free SO_2. The explanation offered is that these patients had insufficient sulfite oxidase enzyme. When the bound SO_2 got into the gut at the higher pH, it was hydrolysed. It entered the bloodstream and caused a reaction. This evidence was sufficient for the industry to accept the ruling of the FDA without further hesitation. As far as this author knows, no one has clearly shown that ingestion of wine made commercially caused a death directly related to SO_2 in the wine. Nevertheless, asthmatics should be aware of the danger. The concentrated form (liquid as a compressed gas) should be handled with utmost care. It can blind a person as easily as any other acid if it gets into their eyes. Prolonged breathing of fumes at relatively low levels can permanently damage lung tissue and contribute to emphysema. When working with SO_2, either solution or liquified gas, the proper protective glasses, clothing and mask should be worn.

Wedzicha (1984) reviewed the toxicological work done with rats. He noted free SO_2 was absorbed directly through the stomach wall. It was bound in the plasma as S-sulfonates, mainly to disulfide containing enzymes.

The ADI (adult daily intake) is recommended not to exceed 0.5-0.8 mmol/day or 32-51 mg/day. Wine probably is the product most likely to add to the daily SO_2 intake. If a person had one-half of a bottle of wine with an average SO_2 content of 100 mg/L, they would probably be close to the recommended ADI.

Sorbic Acid

Sorbic acid at the levels used in wine is only a yeast inhibitor. Around 500 mg/L is the amount required for killing the yeast. It is relatively inactive towards bacteria. On the other hand, bacteria can react with sorbic acid to form a rather disagreeable compound. This can happen in wines that are bottled without sufficient SO_2 to prevent bacterial growth. Crowell and Guymon (1975) determined the reactions shown below:

$$CH_3{-}CH = CH{-}CH = CH{-}COOH + C_2H_5OH \xrightarrow{H^+} CH_3{-}CH = CH{-}CH = CH{-}\overset{\overset{O}{|}}{C}{-}O{-}C_2{-}H_5$$

Sorbic acid ethyl sorbate

↓ lactic acid bacteria

$$CH_3{-}CH = CH{-}CH = CH{-}CH_2OH \xrightarrow{\text{Ethanol}} CH_3{-}CH = CH{-}CH = CH{-}CH_2O{-}C_2H_5$$

sorbyl alcohol 2,4-hexene-1-ethyl ether

↓ rearrangement, H+

$$CH_2 = CH{-}CH = CH{-}CHOH{-}CH_3 \xrightarrow{\text{Ethanol}} CH_2 = CH{-}CH = CH{-}\underset{}{\overset{\overset{O-C_2H_5}{|}}{CH}}{-}CH_3$$

3,5-hexadien-2-ol 2-ethoxyhex-3,5-diene

They concluded the major disagreeable odor (geranium-like) was the 2-ethoxyhex-3,5-diene. It was suggested the other two alcohols and the ether might contribute to a small extent. The ethyl sorbate is the normal odor associated with wines treated with sorbic acid. The development of the ester is relatively slow. The sensitivity to it varies from person to person. Sorbic acid itself is quite odorless when put onto water solution. Sorbic acid actually is very insoluble and it is used commercially as the potassium salt which is soluble at wine pHs.

There has been one report by Heintz (1976) that sorbic acid would form a reaction product with SO_2.

$$CH_3{-}CH = CH{-}CH = CH{-}COOH \xrightarrow{HSO_3^-} CH_3{-}\underset{\displaystyle SO_3}{\overset{\displaystyle H}{C}}{-}CH_2{-}CH = CH{-}COOH$$

Parrrish and Carroll (1988) show some antagonistic reactions between sorbic acid and SO_2.

Sorbic acid is regarded as a very safe additive. The level approved for use in foods and beverages is 300 mg/L. This exceeds what is commonly used in wine. The obvious odor problems and cost encourage minimum use.

Dimethyl Dicarbonate

Dimethyl dicarbonate (DMDC) is a compound that has been available for use for 10 years or more. It was approved by the FDA late in 1988. This compound was developed after its analog diethyl dicarbonate (DEDC) was banned from use. The reason was that the carcinogen ethyl carbamate was formed. This compound (DEDC) was used for almost 10 years as a yeast sterilant. It was very effective and had replaced sorbic acid completely as a fungicide of choice in wine.

$$C_2H_5{-}O{-}\overset{\displaystyle O}{\overset{\|}{C}}{-}O{-}\overset{\displaystyle O}{\overset{\|}{C}}{-}O{-}C_2H_5 + NH_3 \longrightarrow C_2H_5{-}O{-}\overset{\displaystyle O}{\overset{\|}{C}}{-}NH_2 + CO_2 + C_2H_5OH$$

The amount of ethyl carbamate that formed was small and only added a few parts per billion to the normal level found in wine. The formation was dependent on the pH as the reactive form of the ammonia was the unionized form. However, despite some vigorous protests by the industry, the product was banned from use. The reason that the methyl analog was not originally used commercially was the concern for the methanol formed by hydrolysis. As with the DEDC, most of the product is hydrolysed.

$$CH_3-O-\overset{O}{\overset{\|}{C}}-O-\overset{O}{\overset{\|}{C}}-C-CH_3 + H_2O \longrightarrow 2CH_3OH + 2CO_2$$

One hundred mg/L of DMDC will give about 46 mg/L CH_3OH. Considering legal limits and health-related matters, this is not a concern. The methyl carbamate is not considered a carcinogen.

The DMDC has a melting point of 16°C and a density of about 1.25. The usual method of addition is to atomize it into the wine-bottling line with a special proportional pump. For smaller operations, after dissolved into a small amount of absolute ethanol, it is added immediately to the wine tank. The wine is mixed thoroughly and the bottling started immediately. The bottling line will be sterilized for yeast providing there is minimum delay in the operation. The half-life of DMDC is very short. Figure 10-1 gives the half-life. Other compounds can form. The only one of significance is the ethyl methyl carbonate discussed in Chapter 7. Peterson and Ough (1979) investigated the other alcohol side-products. The amounts expected to be produced from normal wine and normal levels of DMDC were less than 0.1 mg/L.

The DMDC has some undesirable features. It is a strong lachrymator and will burn the skin. It is very reactive and should not be taken internally. Once it reacts, however, the side-products are safe at the levels formed. Care in handling should be stressed.

The optimum conditions for use of DMDC were investigated by Porter and Ough (1982). The most effective temperature was found to be about 20°C. The two competitive reactions are the hydrolysis of the DMDC and the reaction of DMDC with the enzymes. Other factors such as ethanol and pH are also important. The postulated reaction of the enzyme is seen in Figure 10-2. The DMDC has no effect on enzymes with metal active sites.

The pure material breaks down autocatalytically. If a small amount of water or ethanol gets into the pure material, it breaks down very rapidly even at refrigerator storage temperature. The

FIGURE 10-1. The half-life of dimethyl dicarbonate as it is affected by temperature in 14.6% ethanol in water.

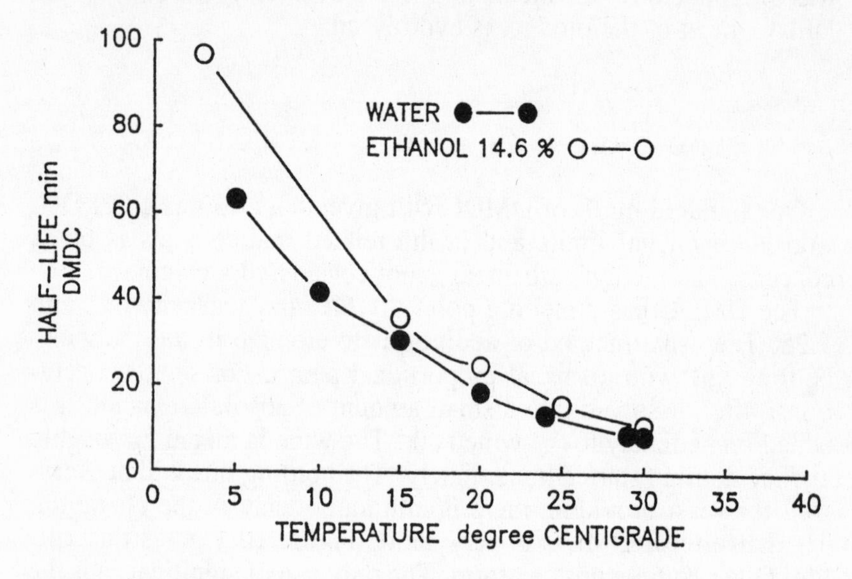

pure material can be analyzed by reacting with a known excess of morpholine and back titrating with methanolic hydrochloric acid. This should be done on any new material received or on one held for further use. The evidence of CO_2 effervescence is a good indicator that the compound has begun to break down.

Ascorbic Acid

The use of ascorbic acid or its isomer, erythorbic acid, to scavenge oxygen from wine has been practiced for many years. The pros and cons of its use have been discussed by many.

The reaction shown at the top of page 281 is suggested as the main one. If iron or copper is present, autocatalytic reactions can go on. Obviously, hydrogen peroxide must not be present in the juice or wine. That can be accomplished by the action of SO_2 with the peroxide to form sulfate and water. Studies (Bertheau, 1988) are not convincing that the use of ascorbic acid is merited.

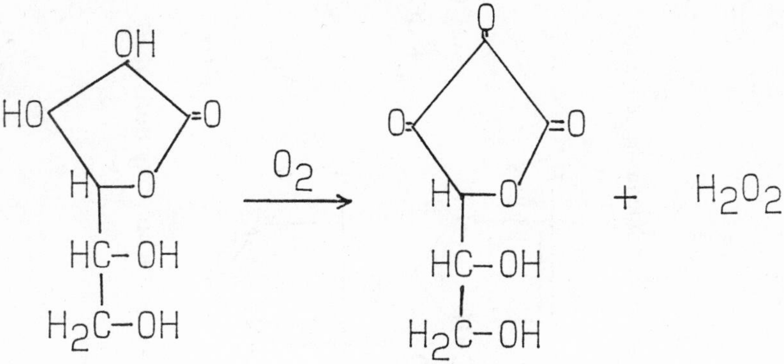

OH

HO =O

O_2 →

H —O

HC−OH

H_2C−OH

O

O= =O

H —O + H_2O_2

HC−OH

H_2C−OH

Ascorbic Acid Dehydroascorbic Acid

Ascorbic acid, or a cheaper isomer substitute erythorbic acid, is used extensively in Australia for white table wines to scavenge oxygen (Rankine, 1987). He suggests as long as 15-35 mg/L of free SO_2 is present, its use is satisfactory. His recommendations for ascorbic acid are 50-100 mg/L at the crusher and maintain that level until bottling. The use of ascorbic acid will rapidly reduce the measurable oxygen. Whether the rapid removal of the oxygen is preferable to the natural reduction by wine components is the question. In one case, the H_2O_2 produced may cause browning and oxidative damage by side-reactions with other than SO_2. In the other case, the unreacted oxygen may also cause oxidation. It seems that the slower production of H_2O_2 by the natural wine component reactions would be preferable to a relatively rapid formation of H_2O_2 within a short time and at higher concentrations by the use of ascorbic acid. In either case, if sufficient SO_2 is present, the peroxide oxidation will be fairly well controlled.

ILLEGAL OR UNUSED YEAST INHIBITORS

Many materials have been used to inhibit spoilage in grape juice and wine. Many are very effective but have adverse effects or potential hazards can be foreseen. The main reason for disapproval of

FIGURE 10-2. The effective reaction of dimethyl dicarbonate with enzymes containing histidine in the active site.

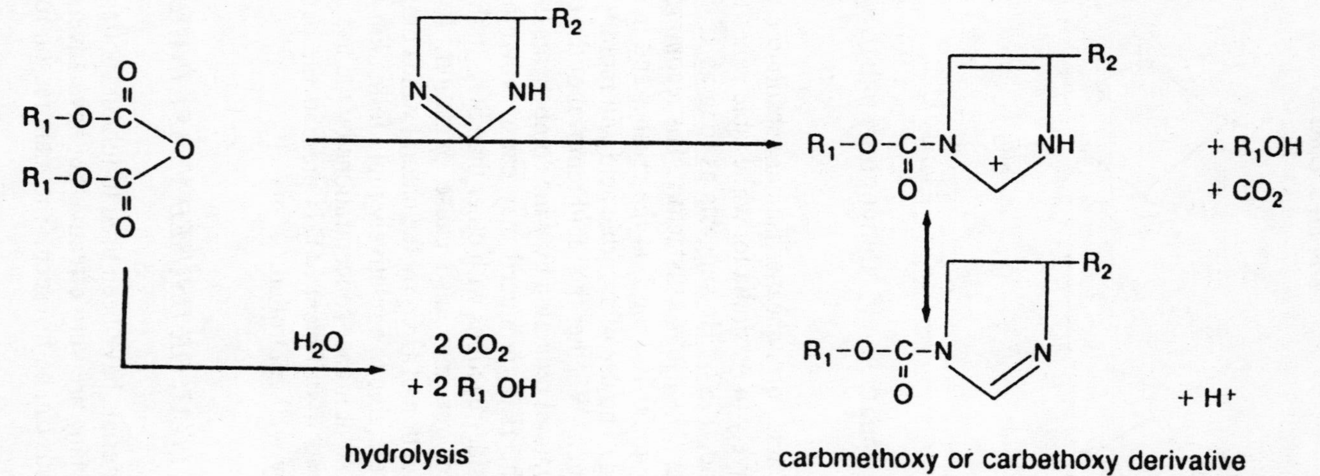

these additives' use is their effective level is too close to their toxic level for humans. Errors in addition of tenfold can put some of them into the near lethal area. Many have been used effectively previously. Some were used even fairly recently during shortages of SO_2 during World War II. Most of these compounds are odorless and tasteless at the toxic levels, further increasing the danger to the consumer.

A mixture made up of benzyl alcohol, benzylbromacetate and dibenzyl ether was used illegally in Austria in a product called "Sterile."

The toxic levels of several yeast-inhibiting wine additives were summarized by Flak and Schaber (1988). In Table 10-2, the levels expected to be found for inhibition of *S. cerevisiae* in wine and the rat LD_{50} levels are shown. They could separate and quantify most of these inhibitors in one run with HPLC using diode array detection.

Robertson (1983), using a very sensitive fluorometric method, found that salicylic acid was a normal component of wine varying from 40 to 70μg/kg. It is found in carbonic maceration wines.

5-Nitrofuryl Acrylic Acid

5-Nitrofuryl acrylic acid is used legally in the Eastern Bloc countries and in Russia.

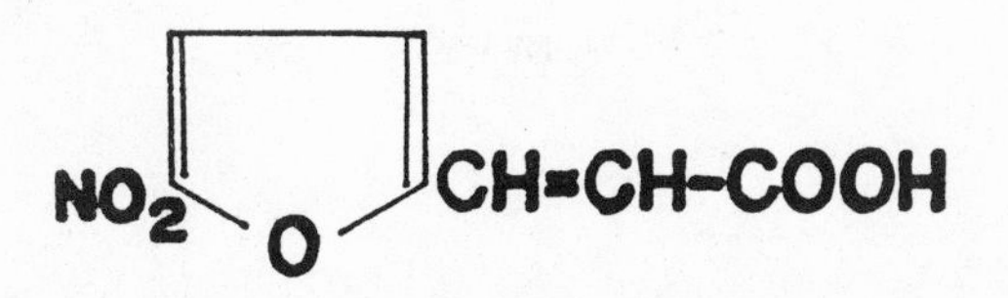

Effective levels are between 1 and 5 mg/L. Amounts found in wines from these countries vary from 0-8.7 mg/L and average between 1 and 2 mg/L. It is illegal in the USA and European Economic Community for two reasons: (1) the amount needed exceeds the toxic safety levels normally used, and (2) a compound of this nature could form a nitrosamine.

Table 10-2. Additives for Yeast Inhibition, Range of Use Levels and Rat Toxicity Data.

Preservative	Level for effective inhibition mg/L	Actual oral toxicity rat L.D.$_{50}$ g/kg
Sorbic acid	50-200	10.5 ± 1.96
Benzoic acid[1]	500-2000	1.7 – 3.7
p-Hydroxybenzoic ethyl ester[2]	800	6
p-Hydroxybenzoic n-propyl ester	400	8
p-Hydroxybenzoic methyl ester	>800	8
Salicylic acid	500	1.1 – 1.6
Chloroacetic acid	-	0.076
p-Chlorobenzoic acid	100-300	-
5-Nitrofuryl acrylic acid	5-10	-
Sodium azide	-	1 – 3 mg/kg[3]
Pimaricin	4 – 10	1.5 – 4.7

[1]Benzoates must be declared on the label. It is relatively ineffective in wine and has adverse taste effects.

[2]These esters vary in effectiveness. The longer the chain length of the alkane portion the more effective it is, but the less soluble. Some have noted an anesthetic effect in the mouth.

[3]Limit for human survival. Toxicity effects appear at 0.01 mg/kg.

Sodium Azide

Sodium azide, NaN_3, is a very potent compound. As little as 1-3 mg/kg of body weight is sufficient to kill and as little as 0.01 mg/kg causes a detectable drop in blood pressure. Unscrupulous wine merchants of France have been caught using the compound and boatloads of wine have been confiscated. The sodium azide blocks the activity of ATP, a necessary energy-giving compound in the body.

Halogenated Acids

The halogenated acids of acetic acid and their ethyl esters were used as preservatives until after about 1940. These are either the chloro- or bromo-substituted acids:

$$CH_2Cl-COOH$$

$$CH_2Br-COOH$$

Their ethyl esters were also used. As with many of these other compounds, the safety factor for use in foods was insufficient.

Pimaricin

Pimaricin is fungicide used for topical treatment of yeast infections on humans. It is classified as an antibiotic in the USA. It is not approved for use in foods and beverages. This compound inhibits yeast at 5-10 mg/L. It does not kill yeasts at these levels. The yeast can overcome the inhibitive effects. It is photodecomposed. It is also difficult to solubilize for use in a concentrated form. It has been tested for use in Europe with implied success. Figure 10-3 gives the rather complicated structure of this compound.

COMPOUNDS ADDED FOR FLAVOR OR COMPOSITIONAL CHANGE

Diethylene Glycol and Other Glycols

Diethylene glycol ($HOCH_2-CH_2-O-CH_2-CH_2OH$) is a toxic substance. It has the property of enhancing the mouth feel of a wine. It gives the taster the impression of enhanced sweetness and flavor. The wine appears to have a heavier body. A few years ago, many wines from Europe were found to contain this compound illegally. The losses from the seized wines were significant. This substance has a LD_{50} for rats of 20.8 g/kg.

Its toxic effects are similar to that of ethylene glycol (CH_2OH-CH_2OH) used as an antifreeze compound. The LD_{50} for rats of this compound is 8.5 g/kg. Propylene glycol ($CH_3-CH_2OH-CH_2OH$)

FIGURE 10-3. The chemical formula of primaricin (also called tennecetin, myprozine, natamycin and pimafucin).

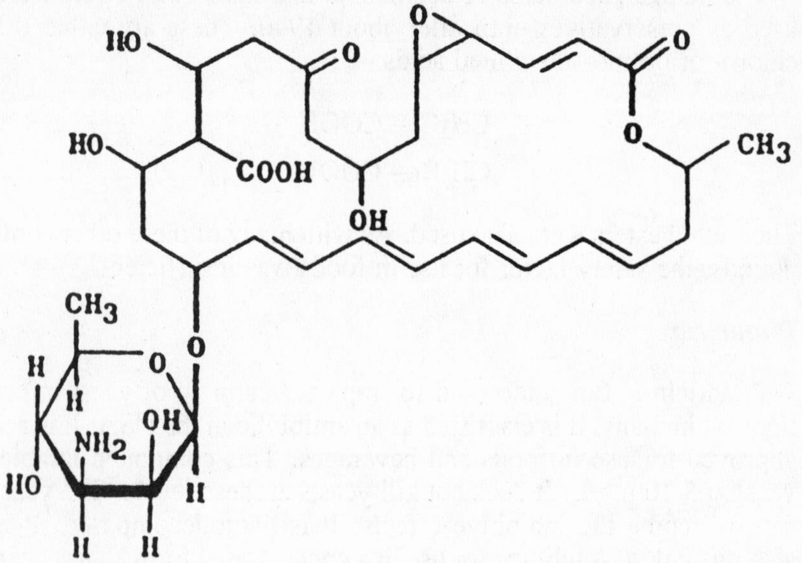

is somewhat less toxic: LD_{50} for rats is 30g/kg. This is a common chemical used in heat exchangers in wineries.

Methanol

The substitution of methanol for ethanol in wines is illegal in all countries. However, it is the main cause of deaths associated with wines. Methanol can be obtained by the distillation of wood (hence, the name wood alcohol). It is cheaper and easier to obtain than ethanol. Recently, wines made with insufficient ethanol were fortified with methanol in Italy. Twenty-two people died from this product. People who make illegal artificial wines also use as much methanol as they dare because of the cost savings. Methanol affects the optic nerve and causes blindness. It is not the methanol per se that kills a person but the formaldehyde formed by the oxidation of methanol by the alcohol dehydrogenase.

$$CH_3OH \xrightarrow[\text{alcohol dehydrogenase}]{NAD^+ \rightarrow NADH} H_2C=O$$

The treatment for a person with an overdose of methanol is to give them all the ethanol they can safely consume. The ethanol saturates the enzyme and slows down the amount of formaldehyde being produced and allows the body a chance to detoxify itself.

For those interested in some historical background on wine scandals of the past, suggested reading is Hallgarten (1986).

Special Flavorants

In the past, many potent flavoring agents were added to wines. The purpose probably was mainly to cover up the poor quality wine that was being produced. Not too many years ago, wines had excessive spoilage, were oxidized and fermented very hot. This was due to a lack of understanding or means to control these problems. However, many of the herbs and materials used were not very well understood concerning the possible adverse effects on humans. Among these compounds are α- and β-thujone (carcinogens), safrole (toxic substances) and woodruff used in many wines (it contains coumarin, a fairly toxic substance). Their use in the USA is prohibited or strongly controlled as to amounts by the BATF.

NATURAL INHIBITORS

There are some natural bacteria and yeast inhibitors found in grapes and wines. In a review article (Beuchat and Golden, 1989), a few of these were mentioned. The medium chain fatty acids and their esters have, for over 30 years, been known to inhibit microorganisms. Although some doubt remains, it is the consensus of opinion that these compounds cause cell membrane leakage. This interrupts the electron transport chain associated with ATP regeneration. It is well documented that the effects of these fatty acids can be overcome by various antagonists including certain protein, starches and sterols. Anthocyanin glycosides are also inhibitory. These chelate Mg^{++} and Ca^{++} to prevent growth of the microorganism. Other inhibitory phenol compounds are vanillin, cinnamic aldehyde, p-coumaric acid, ferulic acid, caffeic acid, pro-anthocyanins and flavonols. These are forbidden or controlled for use by the BATF. Also, the ethanol content, as well as the low pH, of the wine makes a very inhospitable medium for microorganisms. It is one of the few foods or beverages in which absolutely no pathanogenic bacteria or yeast will survive.

SPECIAL NATURAL WINES AND COOLERS

A category of beverages usually containing some grape wine is special natural wines and coolers. The former is usually prepared from an apple or pear juice base (very neutral). A blend of natural flavors is added. The rule on the flavors is they must come from a natural source—not be synthesized. Coolers have a wine/natural fruit juice mix and have flavors added. These formulas of added flavors are approved for use by the regulating agency (BATF).

FRAUD DETECTION

Fruit juices are traded all over the world. When there is a shortage of apple juice in California, apple juice concentrate may be purchased from as far away as South Africa. The detection of fraudulent shipments has become a fine science. Anyone wishing to substitute apple juice for grape or pear juices can be caught rather easily. It is more expensive and difficult to catch people who add sugar to wines illegally but that can also be done. Consider the number of chemical components of grapes, apples and pears. Table 10-3 gives the compounds that may or may not be overlapping in amounts.

There is extensive data on methods to detect adulteration in Nagy et al. (1988). Ratios of carbohydrate to many of these components can be used to determine if these ratios are out of their normal distribution area. This is especially good for detecting the adulteration of the juice by sugar syrups.

Sugar addition from a source such as corn or cane (C_4 pathway or Hatch-Slack) can be detected by a measurement of the $^{13}C/^{12}C$ ratio. Beet sugar follows the normal Calvin pathway (C_3 plants) of photosynthesis.

Table 10-3. Compounds Which Can Be Used to Separate Pure Juices of Grape, Pear or Apple.

Compound	Apple	Pear	Grape
Proline	L	M	M
Asparagine	M	M	L
Alanine	L	L	M
γ-Aminobutyric acid	L	L	M
Arginine	L	L	M
Tartaric acid	N	N	H
Malic acid	H	M	M
Sorbitol	M	H	L

L = low, M = medium, H = high, and N = not present

There are two different enzymes in plants that fix the CO_2. In the Calvin pathway (C_3 plants)

$$CO_2 + \text{ribulose diphosphate} \xrightarrow[\text{carboxylase}]{\text{Ribulose diphosphate}} \text{oxalacetic acid}$$

and in the Hatch-Slack pathway (C_4 plants)

$$CO_2 + \text{phosphoenolpyruvate} \xrightarrow[\text{carboxylase}]{\text{PEP}} \text{oxalacetic acid}$$

The C_3 carbon dioxide fixing enzyme favors the $^{12}CO_2$, and the C_4 enzyme shows no discrimination. Apple, grape and pear are all C_3 plants. The addition of relatively small amounts of C_4 plant sugar (corn syrup or cane invert sugar) is readily detectable by the $^{13}C/^{12}C$ ratio measurements. Beet sugar is from a C_3 plant and cannot be so detected. Work (Donner et al., 1987) in orange juice has indicated differences exist in the carbon attached D/H and $^{18}0/^{16}0$ ratios of the sugars when compared to adulterants. This offers another possible method for screening for adulteration.

All of the possible ways to determine if, for example, pear, apple or grape juice have been mixed, results in a statistical exercise. The proof of mingling of juices is seldom absolute but a probability. If one purchases juices, it is best to evaluate to see if the juice meets sensory and analytical requirements. Purchase with a statement from the seller that the juice is what he/she claimed it to be. When the juice is received, repeat the analytical tests. If values do not coincide with original sample, reject the juice. The analytical tests should be some of the ones shown in Table 10-3 if possible (proline, arginine, tartaric acid, malic acid and sorbitol). If it is supposed to be apple, proline will be very low (< 50 mg/L) and tartaric acid will not be present. If apple and grape have been mixed, proline and tartrate will have relatively large values. It is possible arginine will be larger than normal. If pear and apple are mixed, it is more difficult to differentiate. Proline and sorbitol may be higher than acceptable. Pear and grape are relatively easy to differentiate. The tartaric acid and sorbitol are the key screening chemicals, with alanine and amino acids as possible aids in detection.

Apple wines were found not to contain acetamides in contrast to grape wines (Stackler and Ough, 1979). The only value in detection of adulteration would be the presence of the acetamides in a fermented juice claimed to be 100% apple. Again, it would have little value in discrimination of diluted grape juice with apple.

Chapter 11

The Home Winery

The amount of money spent for a home winery can vary from practically nothing to thousands of dollars. The quality of wine may not increase in proportion to the amount of money spent. The main goals that should be attained are cleanliness, good practices and proper physical conditions. Winemaking is an endeavor favored by constant changes in the starting material and many variable that can affect the wine. Continued learning to control conditions is desirable.

Commercial wineries come in all sizes and costs. This chapter is aimed solely for the amateur but some aspects apply to commercial wineries. For a commercial winery, it is vital to have a qualified design engineer. He should have extensive experience in winery design. The sizing of equipment for refrigeration, materials handling, electrical needs and such concerns as waste removal, future expansion, etc., all demand a great deal of knowledge. This is not gained from books but must be purchased. For home wineries, much less expertise is required.

CRUSHING AND FERMENTING BUILDING

A basement or a barn can suffice well for the home winery. What is needed, basically, is an area where the grapes can be crushed, fermented and pressed. Ease of cleanup and maintenance of proper fermenting and storage temperatures should be considered. Convenience should be considered also. Locating the building on a hillside has some advantages. These are: ease of movement of material and possible insulation by partial or full excavation for a storage area.

The building should have a concrete floor with sufficient slant so water will run off to drains and can easily be cleaned of juice, stems and grapes. There should be no equipment or tanks located so it is difficult to clean under as it becomes a source of contamination. The crushing and fermentation equipment should be shaded if possible. Bees can be a real problem when handling boxes of grapes after they have been stacked in the sun. However, the problem is greatly minimized if the boxes are in shade. Also, the fruit does not get as hot. The building need not have walls for this part of the operation. If the grapes come in gondolas and are dumped into the crushing pit, then shade is not a requirement.

Figure 11-1 shows a rough schematic of a crushing, pressing and fermenting arrangement using the advantage of a hillside location. The fruit starts at the top and works its way down the levels with a minimum of handling.

If a hillside location is not available, then must pumps have to be used if more than a minimum of fruit is being handled. For most home winemakers, the crushed material is handled by hand with buckets.

FACILITIES

If a special building is constructed, then be sure ample water and electricity are put in when it is built. Determine what your total amperage needs would be if all the equipment you anticipate ever having was going at once and then triple it at least. Invariably the need grows with time. Locate outlets in convenient locations—do not depend on long cords on the equipment. Local planning commission or city or county engineers can give the details on electrical code requirements. Pipe in sufficient water.

EQUIPMENT

The type of refrigeration required for fermenting white wines depends on the tanks. One hundred to 300 gal stainless steel tanks are suggested for home wineries. These should have removable cooling plates or jackets. A closed top with a 20-inch diameter lid is necessary. A small refrigeration unit (3-5 H.P.) with a tube-in-shell heat

FIGURE 11-1. Schematic of a hillside, minimum handling table wine arrangement. A. Crusher-stemmer, B. Stem removal cart, C. Juice settling tank (refrigerated), D. Refrigerated skin contact tank, E. Press, F. Juice fermenting tank, G. Refrigeration unit, and H. Pump.

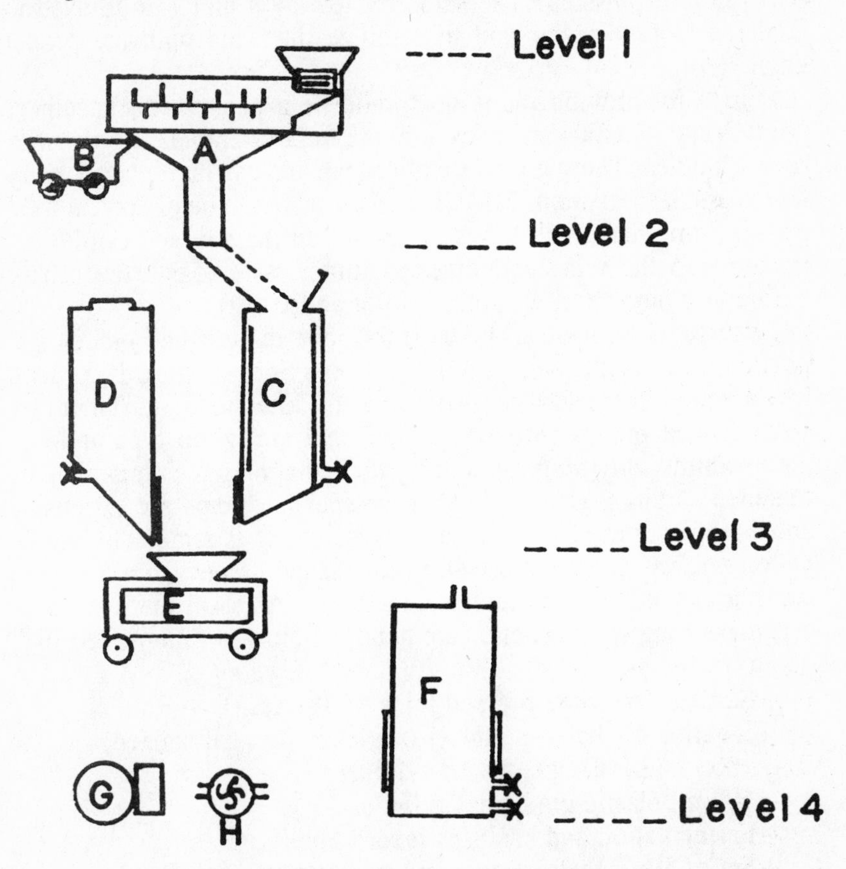

exchanger is required. A pump and hose arrangement to move chilled propylene glycol (or water) to the cooling plate or jacket of the tank is satisfactory. Depending on the initial juice temperature, the tank size, the rate of fermentation and the control temperature desired, several tanks can be controlled within certain limits. The system can be as elaborate or as simple as one wants to make. White wines should be fermented between 15 to 18°C for good results. Red wine fermentations have a significant refrigeration demand until they are pressed, then the fermenting juice should be treated as white wine. If larger tanks are used, then the temperature of the red fermenting musts should be kept around 24-30°C (75-85°F) before pressing. The same refrigeration unit can be used to cool the storage cellar and to stabilize the wine with the proper engineering.

Pumps for moving the wine should be a positive displacement type. Very satisfactory ones are the rubber impeller type with bronze bodies. These are relatively cheap and can be replaced every few years or so economically. Hoses used should be of special food grade. Some have material incorporated in them which can be extracted into the wine and give it a bitter taste. Taste the rubber before you buy it (see Chapter 7 for a better test).

Centrifugal pumps can be used to move the crushed grapes and juice. At least 3-inch and preferably 4-inch lines are best. If the fruit is not juicy, there is a tendency for the lines to clog. There are special must pumps (Mono® pumps) that are worm type and can pump almost anything. Also piston or diaphragm pumps can be obtained to pump must. All of these special pumps are expensive and hardly justified for a small operation. It is simple to use 3-gallon buckets and hand transfer the crushed grapes from a mobile receiving tank to the fermenter or the press. An assortment of buckets, measuring devices, etc., are handy. Below is a suggested list:

- Screens for white pressed juice (8-10 mesh)
- 3-gallon buckets – stainless or plastic (not galvanized)
- 1000 ml plastic graduated cylinder
- 100 ml plastic graduated cylinder
- Fermentation and analyses record book
- Set of °Brix hydrometers and thermometer – with rack
- Plastic hydrometer cylinder with overflow
- 7-foot length of food grade rubber hose 3/8″ inside diameters
- Mobile tanks to hold about 80 gals crushed grapes
- Wooden or metal plunger for punching down caps and for mixing in SO_2 and yeast
- Shovel for pomace removal
- Selection of sample bottles and 1/2 gal, gallon and large glass containers

–Bung starter
–Assortment of brushes and broom and squeegee
–Tasting book for comments and observations
–The filtration equipment has already been described

If smaller lots are desired, then 12-gal Pyrex carboys with fermentation locks are very useful. This would require some room refrigeration (Figure 11-2).

CELLAR

The wine is fermented, racked, malolactic fermented if desired, and stabilized. The wine then goes into barrels or, in the case of some whites, into stainless steel tanks or glass bottles for aging.

An ideal cellar storage temperature is about 15°C. The floor should be concrete and the walls and ceiling should be insulated. If the area climate has cool nighttime temperatures, an arrangement to use the cool night air to maintain the desired temperature is economical. This involves having doors which open at each end of the building in the direction the wind blows. If, for security reasons, this is not desirable, then an attic exhaust fan with louvered opening at the base of the building is satisfactory. Lighting should be considered. Skylights do not take up less valuable wall space.

If a home winemaker makes 300 gals of wine/year of which half is red and half is white, this requires three 55-gal stainless steel barrels for aging of the white wine and three 50-gal oak barrels for aging the red wine. If the red wine is kept in the barrel for more than one year, there should be room for at least 10 or 12 50-gal oak barrels. Occasionally, you might put some white wine into a barrel for short-term aging. The barrels should be on permanent racks which are sturdy and have good clearance off the floor for maintenance of cleanliness.

Wine for filling the barrels should be shelved by the barrel rack and tagged with lot number. The barrel also should be tagged with a permanent identification number. If anything happens, such as bacterial spoilage, to a wine while in the barrel, the barrel can be later identified. Barrels for one reason or another sometimes have to be discarded.

FIGURE 11-2. Juice fermenting in 12 gal pyrex bottles in a temperature controlled room.

Cleanliness

The fermenting room and cellar should be clean. After each day's crushing and pressing, all the equipment should be thoroughly cleaned. There should be no seeds, skins or grapes on the floors or on the equipment. The presence of these encourages fruit flies which, in turn, can increase the bacteria and wild yeast populations. The best tools for cleaning are water, brushes, brooms, squeegees and effort. Steam cleaning with detergent is very effective. Any spilled juice or wine should be cleaned up immediately. Leaky barrels should be repaired or replaced. Pomace and stems should not be stockpiled by the winery area. They can be carted away to the dump, put back onto the vineyard or taken away. When barrels are filled, all spillage should be washed away and the area brushed off with 1000 mg/L SO_2 solution.

Equipment

Bottling equipment, hoses, filters, pumps and other items used to handle the wines during storage, racking and stabilization should be sterilized. This is done with either a strong SO_2 solution (1000 mg/L) or some appropriate commercial sterilizing compound. The equipment is rinsed with water and allowed to dry before storing. This will help prevent transmission of microbiological diseases between wines if an outbreak occurs.

Mold growths may occur on the concrete floor for some reason. A treatment with a bleaching agent is a rapid way to remove the mold and bring the floor back to its original color. (Be careful to have good ventilation when using chlorine powders or liquids and when using strong SO_2 solutions. Both are toxic and can cause serious damage to the lungs.) Steam or high-pressure solutions of detergents at proper times are recommended. This will negate the need for more drastic treatments.

Bottle Storage Bins

Bottled wine can be binned. This form of storage is an efficient use of space. The disadvantage is that only the top row can be easily used. Planks (2 in $\times$ 12 in) are laid on concrete block supports. The end braces are best walls or very strong, well-braced end pieces. Uprights can be 1 in × 12 in planks. Adjustable shelf strips can be put onto the front and back of each side of a bin for less than full bin usage. Figure 11-3 gives some dimensions and a schematic of a possible arrangement. The position distances depend somewhat on the number of bottles being laid down and the number of different wines each year. Each partition, as shown, is set up to hold 120-130 bottles. Thus it will take two bins to hold the bottles for a 50-gal barrel. It is better to have them no bigger than this as you will probably have 10-gal size lots, or even less occasionally. Then the adjustable shelves can be used. You can store the wine in larger bottles also. These can be purchased in 1-, 2-, 3- and 5-gal sizes. They should be filled full and closed with inert stoppers (rubber

FIGURE 11-3. Schematic front view of storage bins for bottled wine.

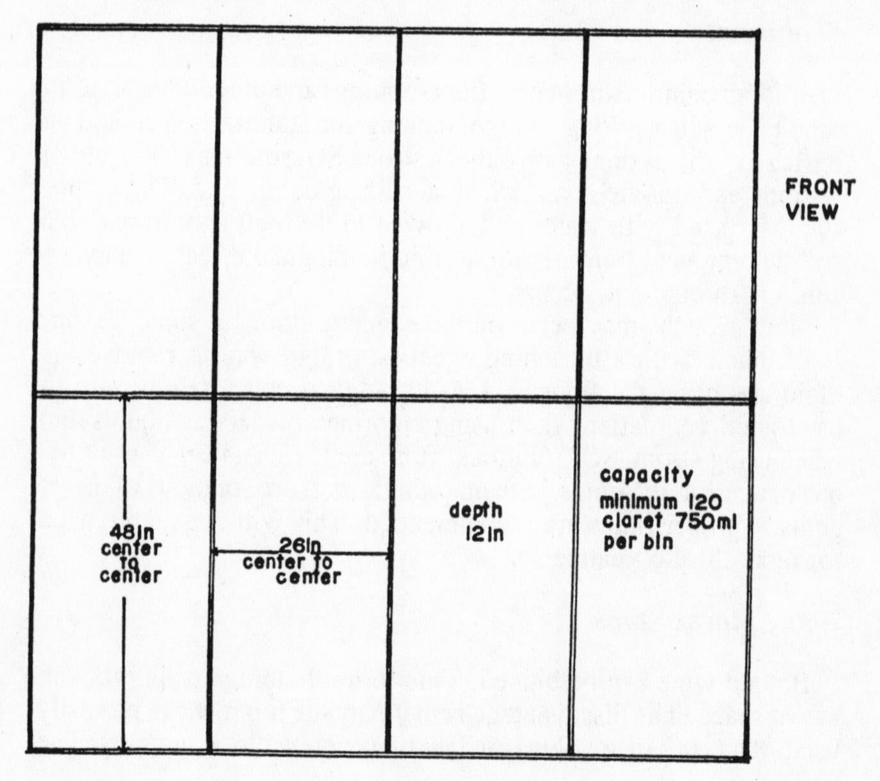

stoppers, Saran® covered). When turned on their sides, the stoppers should be tied or wired in. Alternatively, screw cap closed one gallon bottles are available. Be sure the wine is microbiologically stable. If not, they can push out the closures or explode. In either case, a disagreeable cleanup is necessary. The temperature of storage of the wines is ideally between 13 and 18°C. Holding a reasonably constant temperature is desirable. This prevents the corks from being pushed out by pressure changes. The warmer the temperature, the faster the wine ages.

LABORATORY

A small space with a sink, electricity and gas facilities should be set aside for the laboratory and tasting area. This should be a sheltered area. The kitchen of your house is ideal. However, the use depends somewhat on the commitment of other household members to the wine project. Tasting of the wines should be done in an odor-free area.

Also, near the cellar, you need a workbench with a set of tools for making minor repairs. The extent of this depends on the bent of the individual.

References

Alenso, E., I. Estrella and E. Revilla (1986) Presence of quercitin-3-O-glucuronoside in Spanish table wine. J. Sci. Food Agric. *37*, 1118-1120.

Allen, M. S., M. J. Lacey, R. L. N. Harris and W. V. Brown (1988) Sauvignon blanc varietal aroma. Aust. Grapegrower Winemaker (*292*), 51, 53-54, 56.

Amerine, M. A. (1958) Acetaldehyde formation in submerged cultures of *Saccharomyces beticus*. Appl. Microbiol. *6*, 158-168.

Amerine, M. A. and A. J. Winkler (1963) California wine grapes: Composition and quality of their musts and wines. University of California. Calif. Agric. Expt. Sta. Bull. *794*, 1-83.

Amon, J. M. and R. F. Simpson (1986) Wine corks: A review of the incidence of cork related problems and means for their avoidance. Aust. Grapegrower Winemaker (*268*), 63-64, 66-68, 71-72, 75-77, 79-80.

Anonymous (1985) Ultrafiltration et microfiltration tangentielle in oenology. Proceedings Journées Techniques 23-24 Jan. 1985. 250 p. Inst. Tech. Vigne Vin, Paris.

Astley, C. G. and B. C. Rankine (1985) Functions of cleanness and sanitisers in wineries. Aust. Grapegrower Winemaker (*256*), 43, 45.

Augustyn, O. P. H., A. Rapp and C. J. Van Wyk (1982) Some volatile components of *Vitis vinifera* L. Cv. Sauvignon blanc. S. Afr. J. Enol. Vitic. *3*, 53-60.

Auriol, P., S. Tulasi, G. Goma and P. Strehaiano (1987) Des acidification par *Schizosaccharomyces*. Étude sur les aptitudes des differrentes souches. Approche cinetique sur la degradation de l'acide malique. Rev. Franç. Oenol. *27*(108), 37-42.

Bach, H. P. (1988) Der Einfluss verschiedener Verschlusse auf den Wein während der Lagerung in Abhangigkeit von Fullverfahren und der Lagermethode. 2. Mitteilung: Der Einfluss auf die Sensorik. Wein-Wissen. *43*, 199-213.

Baker, G. A., M. A. Amerine and E. B. Roessler (1952) Theory and application of fractional blending systems. Hilgardia *21*, 383-409.

Baumes, R., C. Bayonove, J. M. Barrillere, J. L. Escudier and R. Cordonnier (1988) La macération pelliculair dans la vinification en blanc. Incidence sur la composante volatiles des moûts. Connaiss. Vigne Vin *22*, 209-233.

Baumes, R., C. Bayonove, R. Cordonnier, P. Torres and A. Seguin (1989) Incidence de la macération pelliculaire sur la composante aromatique des vins doux naturels de muscat. Rev. Franç. Oenol. *29*(116), 6-11.

Baumes, R., R. Cordonnier, S. Nitz and F. Drawert (1986) Identification and determination of volatile constituents in wines from different vine cultivars. J. Sci. Food Agric. *37*, 927-943.

Bayonove, C. (1989) Incidences des attaques parasitaires fongiques sur la composante qualitative du raisin et des vins. Rev. Franç. Oenol. *29*(116), 29-39.

Bayonove, C., R. A. Cordonnier and P. Dubois (1975) Etude d'une fraction caractéristique de l'arome du raisin de la variété Cabernet Sauvignon; mise en évidence de la 2-méthoxy-3-isobutylpyrazine. C.R. Acad. Sci. (Paris) Ser. D. *281*, 75-78.

Bearzatto, G. (1986) Wine fining with silica sol. Aust. N. Z. Wine. Ind. J. *1*(2), 39-40.

Beelman, R. B. and R. E. Kunkee (1987) Inducing simultaneous malolactic/alcoholic fermentations. Pract. Winery/Vineyard *8*(2), 44-46, 50-52, 54-56.

Bell, A. A., C. S. Ough and W. M. Kliewer (1979) Effects on must and wine composition, rates of fermentation, and wine quality of nitrogen fertilization of *Vitis vinifera* var. Thompson Seedless grapevines. Am. J. Enol. Vitic. *30*, 124-129.

Bertheau, R. J. (1988) Interactions of ascorbic acid and sulfur dioxide in grape juice and wine. M.S. Thesis. University of California, Davis.

Beuchat, L. R. and D. A. Golden (1989) Antimicrobiols occurring naturally in foods. Food Technol. *43*(1), 134-142.

Blade, W. H. and R. Boulton (1988) Adsorption of protein by bentonite in a model wine solution. Am. J. Enol. Vitic. *39*, 193-199.

Bott, E. (1986) Centrifugal separation of tartrates from wines stabilized by the contact process. Aust. N. Z. Wine Ind. J. *1*(2), 35-38.

Boulton, R. (1980) The general relationship between potassium, sodium and pH on grape juice and wine. Am. J. Enol. Vitic. *31*, 182-186.

Boulton, R. (1986) Experiences in application of centrifuges to wine production in California. p. 46-52., Internl. Symp. Innovations in Enology. May 28, 1986. Stuttgart-Kellesberg. O.I.V., Paris.

Brigaud, G. and B. Medina (1987) Pour un nouveau type de dégustation. Rev. Franç. Oenol. *27*(109), 11-16.

Buchanan, D. (1985) Oak usage in table wine: a different view point. Aust. Grapegrower Winemaker (*253*), 14.

Cabanis, J. C. (1985) L'histamine et sa toxicité. Bull. O.I.V. *58*(656/657), 1009-1015.

Cabras, P., M. Meloni and F. M. Pirisi (1987) Pesticide fate from vine to wine. Rev. Environ. Contamin. Toxicol. *99*, 83-117.

Campbell, I. (1988) *Culture, storage, isolation and identification of yeasts*. p. 1-8. in "Yeast: A Practical Approach." eds. I. Campbell and J. H. Duffus. IRL Press, Oxford.

Chiba, M., A. W. Brown and D. Danic (1987) Inhibition of yeast respiration and fermentation by Benomyl, Carbendozim, Isocyanates and other fungicidal chemicals. Can. J. Microbiol. *33*, 157-161.

Ciolfi, G. (1988) Lieviti produttori di SO_2. Vini Ital. *30*(1), 23-26.

Cirillo, U. (1989) Comparison entre les adjuvant de filtration (ADF): A base de perlite et de diatomite. Rev. Franç. Oenol. *29*(116), 42-43.

Colograndе, O., A. Silva and A. Lasoli (1984) Acides amines dans les vins mousseux. Connaiss. Vigne Vin. *18*, 27-84.

Conner, A. J. (1983) The comparative toxicity of vineyard pesticides to wine yeasts. Am. J. Enol. Vitic. *34*, 278-279.

Cooke, G. M. and H. W. Berg (1983) A re-examination of varietal table wine processing practices in California. I. Grape standards, grape and juice treatment, and fermentation. Am. J. Enol. Vitic. *34*, 249-256.

Cordier, B. (1987) *Fabrication des barriques*. p. 61-79. in "Le bois et la qualité des vins et des eaux-de-vie." ed. G. Guimberteau. Spec. No. Connaiss. Vigne Vin.

Cordonnier, R., C. Bayonove and T. Baumes (1986) Données recentes sur les precurseurs d'arome du raisin perspectives de leur exploitation en vinification. Rev. Franç. Oenol. *26*(102), 29-41.

Costello, P. (1987) The conduct of MLF under commercial conditions. I. Selecting suitable strains of *Leuconostoc oenos*. Tech. Rev. (Aust. Wine Res. Inst.) (*51*), 17-19.

Costello, P. (1988) The conduct of malolactic fermentation under commercial conditions. II. Use of commercial freeze-dried and frozen starter cultures of *Leuconostoc oenos*. Tech. Rev. (Aust. Wine Res. Inst.) (*51*), 9-12.

Crowell, E. A. and J. F. Guymon (1975) Wine constituents arising from sorbic acid addition, and identification of 2-ethoxy-3,5-diene as a source of geranium-like off odor. Am. J. Enol. Vitic. *26*, 97-102.

D'Agostino, S. (1986) La determinazione dei polifenoli totali nei mosti, mosti muti, mosti concentrati, mosti concentrati rettifificati. Vigne Vini *13*(10), 47-50.

Davis, C. R., D. Wibowo, G. H. Fleet and T. H. Lee (1988) Properties of wine lactic acid bacteria: Their potential enological significance. Am. J. Enol. Vitic. *39*, 137-142.

de la Garza, F. and R. Boulton (1984) The modeling of wine filtrations. Am. J. Enol. Vitic. *35*, 189-195.

Dimitriadis, E. and P. J. Williams (1984) The development and use of a rapid analytical technique for estimation of free potentially volatile monoterpene flavorants of grapes. Am. J. Enol. Vitic. *35*, 66-71.

Donner, L. W., H. O. Ajie, L. O. Steinberg, J. M. Milburn, M. J. De Niro and K. B. Hicks (1987) Detecting sugar beet syrups in orange juice by D/H and $^{18}O/^{16}O$ analysis of sucrose. J. Agric. Food Chem. *35*, 610-612.

Dubernet, M. O. (1974) Aplication de la chromatographie en phase gazeuse a l'étude des sucres et des polyols du vin. These 3rd cycle, Bordeaux.

Dubourdieu, D., Ch. Ollivier and J.-N. Boidron (1986) Incidence des operations prefermentaires sur la composition chimique et les qualités organoleptique des vin blancs secs. Connaiss. Vigne Vin. *20*, 53-76.

Dubourdieu, D., M. Serrano, A. C. Vannier and P. Ribéreau-Gayon (1988) Étude comparee des tests de stabilité proteique. Connaiss. Vigne Vin. *22*, 261-273.

Ducruet, V. (1984) Comparison of headspace volatiles of carbonic maceration and traditional wine. Lebensm.-Wiss. Technol. *17*, 217-221.

Dunsford, P. (1987) Current developments and recent experiences in Champagne. Aust. Grapegrower Winemaker (*284*), 45-46.

Dunsford, P. and R. Boulton (1981a) The kinetics of potassium bitartrate crystalization from table wines. I. Effect of particle size, particle surface area and agitation. Am. J. Enol. Vitic. *32*, 100-105.

Dunsford, P. and R. Boulton (1981b) The kinetics of potassium bitartrate crystalization from table wines. II. Effect of temperature and cultivar. Am. J. Enol. Vitic. *32*, 106-110.

Dupuy, P. (1985) *Rectified concentrated grape must*. 134 p. Commission of European Communities EUR 9441 en.

Edwards, M. A. and M. A. Amerine (1977) The lead content of wines by atomic absorption spectrophotometry using flameless atomization. Am. J. Enol. Vitic. *28*, 239-240.

Edwards, T. L., V. L. Singleton and R. Boulton (1985) Formation of ethyl esters of tartaric acid during wine aging: Chemical and sensory effects. Am. J. Enol. Vitic. *36*, 118-124.

Egbosimba, E. E., E. Linus, C. Okafor and J. C. Slaughter (1988) Control of ammonium uptake from malt extract medium by *Saccharomyces cerevisiae*. J. Inst. Brew. *94*, 249-252.

Escudier, J. L., M. Moutounet and P. Benard (1987) Influence de l'ultrifiltration sur la cinétique de cristallisation du bitartrate de potassium des vins. Rev. Franç. Oenol. *27*(108), 52-57.

Farrer, K. T. H. (1987) *A guide to food additives and contaminants*. 157 p. The Parthenon Publishing Group, Inc. New Jersey.

Fed. Reg. (1990) Revision and recodification of wine regulations. *55* (118), 24974-25034.

Feuillat, M. (1987) *L'élevage des vins blancs de Bourgogne*.

p. 123-141. in "Le bois et la qualité des vins et des eaux-de-vie." ed. G. Guimberteau. Spec. No. Connaiss. Vigne Vin.

Feuillat, M., C. Charpentier, G. Picca and P. Bernard (1988) Production de colloides par les levures dans les vins mousseux élabores selon la méthode champenoise. Rev. Franç. Oenol. *28*(111), 36-45.

Flak, W. and R. Schaber (1988) Die Bestimmung von Konservierungesmitteln in Weinen and anderen Getränken mittels Hochdruckflussigkeitschromatographie. Mitt. Klosterneuburg *38*, 10-16.

Flanzy, C., M. Flanzy and P. Bernard (1987) *La vinification par macération carbonique*. 125 p. INRA, Paris.

Flores, J. H., D. A. Heatherbell, J. C. Hsu and B. T. Watson (1988) Ultrafiltration (UF) of White Riesling juice: Effect of oxidation and pre-UF juice treatment on flux, composition, and stability. Am. J. Enol. Vitic. *39*, 180-187.

Foott, J. H., C. S. Ough and J. A. Wolpert (1989) Rootstock effects on wine grapes. Calif. Agric. *43*(4), 27-30.

Foster, M. D. and M. D. Cox (1984) Winemaking with a decanter centrifuge. Am. J. Enol. Vitic. *35*, 103-106.

Fumi, M. D., G. Trioli, M. G. Colombi and O. Colagrande (1988) Immobilization of *Saccharomyces cerevisiae* in calcium alginate gel and its application to bottle-fermented sparkling wine production. Am. J. Enol. Vitic. *39*, 267-272.

Glories, Y. (1987) *Phenomenes oxydatifs lies a la conservation sous bois*. p. 81-91. in "Le bois et la qualité des vins et des eaux-de-vie." ed. G. Guimberteau. Spec. No. Connaiss. Vigne Vin.

Goniak, O. J. and A. C. Noble (1987) Sensory study of selected volatile sulfur compounds in white wine. Am. J. Enol. Vitic. *38*, 223-227.

Guymon, J. F. (1966) Mutant strains of *Saccharomyces cerevisiae* applied to studies of higher alcohol formation during fermentation. Dev. Ind. Microbiol. 7, 88-96.

Haage, B., S. Krieger and W. P. Hammes (1988) Hemmung der Starterkulturen zur Einleitung des biologischen Säureabbaus durch Spritzmittelbruckstande im Wein. Wein-Wissen. *43*, 261-278.

Harris, R. L. N., M. J. Lacey, W. V. Brown and M. S. Allen

(1987) Determination of 2-methoxy-3-alkypyrazines in wine by gas chromatography/mass spectrometry. Vitis *26*, 201-207.

Heimann, W. (1977) *Thermal processing and the technology of wine*. p. 295-304. in "Physical, Chemical and Biological Changes in Food Caused by Thermal Processing." eds. T. Hoyem and O. Kvale. Applied Science Publisher Ltd., London.

Heintz, K. (1976) Uber die gegenseitige Beeinflussung von Sorbinsäure un schwefliger Säure. Ind. Obst. Gemeueseverwer *61*, 555-556.

Heinzel, M., M. Hagen and C. Bousser (1983) Disinfection des bouchons par l'eau oxygénée activée. Rev. Franç. Oenol. *23*(92), 77-81.

Heresztyn, T. (1986) Formation of substituted tetrahydropyridines by species of *Brettanomyces* and *Lactobacillus* isolated from mousy wines. Am. J. Enol. Vitic. *37*, 127-132.

Heymann, H., A. C. Noble and R. B. Boulton (1986) Analysis of methoxypyrazine in wines. I. Development of a quantitative procedure. J. Agric. Food. Chem. *34*, 268-271.

Hsu, J.-C. and D. A. Heatherbell (1987) Heat-unstable proteins in wine. I. Characterization and removal by bentonite fining and heat treatment. Am. J. Enol. Vitic. *38*, 11-16.

Hsu, J.-C., D. A. Heatherbell, J. H. Flores and B. T. Watson (1987) Heat unstable proteins in grape juice and wine. II. Characterization and removal by ultrafiltration. Am. J. Enol. Vitic. *38*, 17-22.

Iland, P. G. (1984) Studies on the composition of pulp and skin of ripening grape berries. M. Ag. Sc. Thesis. Univ. Adelaide, Australia.

Ingledew, W. M. and R. E. Kunkee (1985) Factors influencing sluggish fermentation in grape juice. Am. J. Enol. Vitic. *36*, 65-76.

Ingledew, W. M., G. A. Magnus and F. W. Sosulski (1987) Influence of oxygen on proline utilization during the wine fermentation. Am. J. Enol. Vitic. *38*, 246-248.

Jacobs, C. J., I. Fourie and H. J. J. van Vuuren (1988) Occurrence and detection of killer yeasts on Chenin blanc grapes and grape skins. S. Afr. J. Enol. Vitic. *9*, 28-31.

Jones, M. and J. S. Pierce (1964) Absorption of amino acids from wort by yeasts. J. Inst. Brew. *70*, 307-315.

Jones, R. S. and C. S. Ough (1985) Variations in the percent ethanol (v/v) per °Brix conversions of wines from different climatic regions. Am. J. Enol. Vitic. *36*, 268-270.

Jordon, A. D. and B. J. Croser (1984) *Determination of grape maturity by aroma/flavour assessment*. p. 261-274. in "Advances in Viticulture and Oenology for Economic Gain." eds. T. H. Lee and T. C. Somers. Proc. 5th Aust. Wine Ind. Tech. Conf. 29 Nov-1 Dec. 1983, Perth.

Joyeux, A. and A. Lonvaud-Funel (1985) Comparison de diverses preparations industrielles de bacteries lactiques réactives pour stimuler la fermentation malolactique. Connaiss. Vigne Vin *19*, 149-159.

Kasimatis, A. N. and E. P. Vilas (1985) Sampling for degrees Brix in vineyard plots. Am. J. Enol. Vitic. *36*, 207-213.

Kishkovskii, Z. N. and I. M. Skurikhin (1976) *Chimiya vina*. 657 p. Pishchepromizdat, Moscow.

Krieger, S., E. de Freen and W. P. Hammes (1986) *Aufuhrung des biologischen Säureabbaue in Wein mit Leuconostoc oenos. Chem*. Mikrobiol. Technol. Lebensm. *10*, 13-18.

Kuensch, U., A. Temperli and K. Mayer (1974) Conversion of arginine to ornithine during malo-lactic fermentation in red Swiss wines. Am. J. Enol. Vitic. *25*, 191-193.

Lafon-Lafourcade, S. and E. Peynaud (1974) Sur l'action antibacteriene de l'anhydride sulfureux sous forme libre et sous forme combinée. Connaiss. Vigne Vin. *8*, 187-203.

Lafon-Lafourcade, S., C. Geneix and P. Ribéreau-Gayon (1984) Inhibition of alcoholic fermentation of grape must by fatty acids produced by yeasts and their elimination by yeast ghosts. Appl. Environ. Microbiol. *47*, 1246-1249.

Langhans, E. and H. A. Schlotter (1987) Die Hefeschönung zur Reduzierung des Kupfer-Gehaltes von Weinen. Wein-Wissen. *42*, 202-210.

Larue, F. and S. Lafon-Lafourcade (1989) *Survival factors in wine fermentations*. p. 193-215. in "Alcohol Toxicity in Yeast and Bacteria." ed. N. van Uden, CRC Press, Boca Raton, Florida.

Lemaresquier, H. (1987) Inter-relationships between strains of *Sac-*

charomyces cerevisiae from the Champagne area and lactic acid bacteria. Letters Appl. Microbiol. *4*, 91-94.

Lemperle, E., E. Kerner and H. Krebs (1988) Glucose-/Fructose-verhaltnis von Traubenbeeren und Weinen. Weinwirtschaft/ Technik (*12*), 19-25.

Lobser, G. J. and R. D. Sanderson (1986) The removal of copper and iron from wine using a chelating resin. S. Afr. J. Enol. Vitic. *7*, 47-51.

Loiseau, G., F. Vezihet, M. Valade, A. Vertes, C. Cuinier and D. Delteil (1987) Contrôle de l'efficacité du levurage par la mise en oeuvre de souches de levures oenologique marquées. Rev. Franç. Oenol. *27*(106), 28-36.

Long, Z. R. and B. Lindblom (1987) Juice oxidation in California Chardonnay. p. 267-271. ed. T. H. Lee, Proc. 6th Aust. Wine Ind. Tech. Conf. 1986, Adelaide.

Lonvaud-Funel, A., A. Joyeux and C. Desens (1988) Inhibition of malolactic fermentation of wines by products of yeast metabolism. J. Sci. Food Agric. *44*, 183-191.

Lonvaud-Funel, A. and P. Ribéreau-Gayon (1977) Le gaz carbonique des vins. II. Aspect technologique. Connaiss. Vigne Vin *11*, 165-182.

Ludemann, A. (1987) Wine clarification with a crossflow microfiltration system. Am. J. Enol. Vitic. *38*, 228-235.

Lui, J. W. R. and J. F. Gallander (1985) Methyl anthranilate content of Ohio concord grapes. J. Food Sci. *50*, 280-282.

Lurton, L. and de J. Guerreau (1988) Étude de la proteolyse des levures de vinification lors de l'élevage d'un vin sur ses lies. Rev. Franç. Oenol. *28*(113), 35-41.

Luthi, H. R., B. Stoyla and J. C. Meyer (1965) Continuous production of flor sherry from New York state wines. Appl. Microbiol. *13*, 11-14.

Maarse, H., L. M. Nijssen, and J. Jetten (1985) *Chloroanisoles: A continuing story*. p. 241-250. in "Topics in Flavour Research." eds. R. G. Berger, S. Nitz and P. Schreier. H. Eichhoun, West Germany.

Maga, J. A. (1989) The contribution of wood to the flavor of alcoholic beverages. Food Rev. Internatl. *5*(1), 39-99.

Marais, J. (1987) Terpene concentrations and wine quality of *Vitis vinifera*. L. cv. Gewürztraminer as affected by grape maturity and cellar practices. Vitis *26*, 231-245.

Marinkovich, V. A. (1983) Allergic reactions attributed to wines. Bull. Soc. Med. Friends of Wine. Sept. p. 3. San Francisco.

Marlatta, G. P., L. G. Favretto and L. Favretto (1986) Cadmium in roadside grapes. J. Sci. Food Agric. *37*, 1091-1096.

Marsal, F., Ch. Sarre, D. Dubourdieu and J.-N. Boidron (1988) Rôle de la levure dans la transformation de certains constituants volatils du bois de chêne au cours de l'élaboration en barrique des vins blancs secs. Connaiss. Vigne Vin *22*, 33-38.

Miller, G. C., J. M. Amon, R. L. Gibson and R. F. Simpson (1985) Loss of wine aroma attributable to protein stabilization with bentonite or ultra filtration. Aust. Grapegrower Winemaker (*256*), 46, 49-50.

Millery, A., B. Duteurtre, J.-P. Boudaille and A. Maujean (1986) Differenciation des trois cépages champenois a partir de l'analyse des acides amines libres des moûts des récoltes 1983 et 1984. Rev. Franç. Oenol. *26*(103), 32-50.

Minárik, E. and O. Jungova (1988) Zur Ausschaltung von hemmwinkenden Substanzen durch. Hefezellwand und Zellulose-Präparate bei der Gärung des Mostes. Wein-Wissen. *43*, 102-106.

Monk, P. R. (1986) Formation, utilization and excretion of hydrogen sulfide by wine yeast. Aust. N. Z. Wine Ind. J. *1*(3), 10-16.

Monk, P. R. and R. J. Storer (1986) The kinetics of yeast growth and sugar utilization in tirage: The influence of different methods of starter culture preparations and inoculation levels. Am. J. Enol. Vitic. *37*, 72-76.

Müller-Spath, H. (1973) *Production of white table wines with minimal* SO_2. p. 51-52. 2nd Wine Industry Technical Conference Proceedings, Tanunda. 7-9 Aug. Australian Wine Research Institute.

Nagel, Ch. W., J. D. Anderson and K. M. Weller (1988) A comparison of the fermentation patterns of six commercial wine yeasts. Vitis *27*, 173-182.

Nagel, Ch. W. and W. R. Graber (1988) Effect of must oxidation on quality of white wines. Am. J. Enol. Vitic. *39*, 1-4.

Nel, L., B. L. Wingfield, L. J. van der Meer and H. J. J. van

Vuuren (1987) Isolation and characterization of *Leuconostoc oenos* bacteriophages from wine and sugar cane. FEMS Microbiol. Letters *44*, 63-67.

Nelson, K. E. and M. S. Nightingale (1959) Studies on the commercial production of natural sweet wines from botrytised grapes. Am. J. Enol. Vitic. *10*, 135-141.

Neradt, F. (1982) Sources of reinfection during cold-sterile bottling of wine. Am. J. Enol. Vitic. *33*, 140-144.

Noble, A. C., R. A. Arnold, J. Buchsenstein, E. J. Leach, J. O. Schmidt and P. M. Stern (1987) Modification of a standard system of wine aroma terminology. Am. J. Enol. Vitic. *38*, 143-146.

O'Brien, K. (1986) Carboxymethyl cellulose and inhibition of tartrate crystalization. Aust. N. Z. Wine Ind. J. *1*(2), 43, 45.

Ooghe, W. and H. Kastelijn (1988) Aminozuurpatroon van druirenmost aangewend voor de bereiding van wijnen van gegarandeerde herkomst. Bel. J. Food Chem. Biotechol. *43*(1), 15-21.

Ough, C. S. (1966a) Fermentation rates of grape juice. II. Effect of initial °Brix, pH and fermentation temperature. Am. J. Enol. Vitic. *17*, 20-26.

Ough, C. S. (1966b) Fermentation rates of grape juice. III. Effects of initial ethyl alcohol, pH and fermentation temperature. Am. J. Enol. Vitic. *17*, 74-81.

Ough, C. S. (1969) Substances extracted during skin contact with white musts. I. General wine composition and quality changes with contact time. Am. J. Enol. Vitic. *20*, 93-100.

Ough, C. S. (1971) Measurement of histamine in California wines. J. Agric. Food Chem. *19*, 241-244.

Ough, C. S. (1984) *Volatile nitrogen compounds in fermented beverages*. p. 199-225. in "Proceedings ALKO Symp. in Flavour Research of Alcoholic Beverages." eds. L. Nykanen and P. Lehtonen. Foundation for Biotechnical Industrial Fermentation Research Vol. 3.

Ough, C. S. (1985) Some effects of temperature and SO_2 on wine during simulated transport and storage. Am. J. Enol. Vitic, *36*, 18-22.

Ough, C. S. (1988) *Determination of sulfur dioxide in grapes and wines*. p. 339-358. In Modern Methods of Plant Analyses. Wine

Analysis. eds. H. L. Linskens and J. F. Jackson, Vol. 6, Springer-Verlag, Berlin.

Ough, C. S. (1989a) The changing California wine industry. J. Sci. Food Agric. *47*, 257-268.

Ough, C. S. (1989b) Unpublished data.

Ough, C. S. and M. A. Amerine (1958) Studies on aldehyde production pressure, oxygen and agitation. Am. J. Enol. Vitic. *9*, 111-122.

Ough, C. S. and M. A. Amerine (1963) Regional, varietal, and type influences on the degree Brix and alcohol relationship of grape musts and wines. Hilgardia *34*, 585-600.

Ough, C. S. and M. A. Amerine (1972) Further studies with submerged flor sherry. Am J. Enol. Vitic. *23*, 128-131.

Ough, C. S. and M. A. Amerine (1988) *Methods for Analysis of Musts and Wines*. 377 p. John Wiley and Sons, Inc., New York.

Ough, C. S., H. W. Berg and M. A. Amerine (1969) Substances extracted during skin contact with white musts. II. Effect of bentonite additions during and after fermentation on wine composition and sensory quality. Am. J. Enol. Vitic. *20*, 101-107.

Ough, C. S. and E. A. Crowell (1979) Pectic enzyme treatment of white grapes: Temperature, variety and skin contact time factors. Am. J. Enol. Vitic. *30*, 22-27.

Ough, C. S. and E. A. Crowell (1987) Use of sulfur dioxide in winemaking. J. Food Sci. *52*, 386-388, 393.

Ough, C. S., E. A. Crowell, R. E. Kunkee, M. R. Vilas and S. Lagier (1987) A study of histamine production by various wine bacteria in model solutions and in wine. J. Food Process. Presev. *12*, 63-70.

Ough, C. S., M. Davenport and K. Joseph (1989) Effects of certain vitamins on growth and fermentation rate of several commercial active dry yeasts. Am. J. Enol. Vitic. *40*, 208-213.

Ough, C. S., D. Fong and M. A. Amerine (1972) Glycerol in wine: Determination and some factors affecting. Am. J. Enol. Vitic. *23*, 1-5.

Ough, C. S. and M. L. Groat (1978) Interaction of soluble solids with yeast strain, fermentation temperature in clarified fermenting musts. Appl. Environ. Microbiol. *35*, 881-885.

Ough, C. S. and A. Kriel (1985) Ammonia concentrations of musts of different grape cultivars and vineyards in the Stellenbosch area. S. Afr. J. Enol. Vitic. *6*, 7-11.

Ough, C. S. and R. E. Kunkee (1968) Fermentation rates of grape juice. V. Biotin content of juice and its effect on alcoholic fermentation rate. Appl. Microbiol. *16*, 572-576.

Ough, C. S. and R. Nagaoka (1984) Effect of cluster thinning and vineyard yields on grape and wine composition and wine quality of Cabernet Sauvignon. Am. J. Enol. Vitic. *35*, 30-34.

Ough, C. S., D. Stevens and G. Jordon (1989) *The effect of hot filling on ethyl carbamate formation in wines*. p. 174-182. International Symp. "Innovations in Wine Technology." June 22-24, 1989, Stuttgart-Killesberg. OIV, Paris.

Oura, E. (1977) Reaction products of yeast fermentation. Process Biochem. *12*(3), 19-21, 35.

Pamment, N. B. (1989) *Overall kinetics and mathematical modeling of ethanol inhibition in yeasts*. p. 1-75. in "Alcohol Toxicity in Yeast and Bacteria." ed. N. van Uden, CRC Press Inc., Boca Raton, Florida.

Parrish, M. E. and D. E. Carroll (1988) Effects of combined antimicrobial agents on fermentation initiation by *Saccharomyces cerevisiae* in a model broth system. J. Food Sci. *53*, 240-242.

Peri, C., M. Riva and P. Decio (1988) Crossflow membrane filtration of wines: Comparison of performance of ultrafiltration, microfiltration and intermediate cut-off membranes. Am. J. Enol. Vitic. *39*, 162-168.

Petering, J., P. Langridge and P. A. Henschke (1988) Fingerprinting wine yeasts. Aust. N. Z. Wine Ind. J. *3*(3), 48-52.

Peterson, T. W. (1978) Reaction of dimethyldicarbonate with higher alcohols and water. M. S. Thesis. University California, Davis.

Peterson, T. W. and C. S. Ough (1979) Dimethyl dicarbonate reaction with higher alcohols. Am. J. Enol. Vitic. *30*, 119-123.

Piergiovanni, L., P. Fava and G. Volonterio (1988) Il vino in barrique. Vini Ital. *30*(5), 17-28.

Pineau, J. (1987) Modification de la composition du raisin par attaque du *Botrytis cinerea*. Rev. Franç. Oenol. *27*(108), 17-20.

Pocock, K. F., C. R. Strauss and T. C. Somers (1984) Ellagic acid deposition in white wines after bottling: A wood derived instability. Aust. Grapegrower Winemaker (*244*), 87.

Porter, L. J. and C. S. Ough (1982) The effects of ethanol, temperatures and dimethyl dicarbonate on viability of *Saccharomyces cerevisiae* Montrachet No. 522 in wine. Am. J. Enol. Vitic. *33*, 222-225.

Postel, W., B. Meier and R. Markert (1986) Einfluss verschiedener Behandlungsstoffe auf den Gehalt des Weins an Megen- und Spürenelementen I. Bentonit. Mitt. Klosterneuburg *36*, 20-27.

Postel, W., B. Meier and R. Markert (1987) Einfluss verschiedener Behandlungsstoffe auf den Gehalt des Weines an Mengen- und Spürenelementen. III. Kieslegur und Perlit. Mitt. Klosterneuburg *37*, 219-226.

Radler, F. (1986) Das in den USA zugelassene Säureungsmittel Fumarsäure and seine Wirkung–Übersicht. Wein-Wissen. *41*, 47-50.

Radler, F. (1987) Wine: Starter cultures and new developments. Wine East. *15*(3), 12-17.

Radler, F. and C. Knoll (1988) Die Bildung von Killertoxin und die Beeinflussung der Gärung durch *Apiculatus*–Hefen. Vitis *27*, 111-132.

Radler, F. and M. Schmitt (1987) Killer toxins of yeasts: Inhibitors of fermentation and their adsorption. J. Food Protect. *50*, 234-238.

Radler, F. and J. Zorg (1986) Characterization of the enzyme involved in formation of 2-butanol from meso-2,3-butanediol from lactic acid bacteria. Am. J. Enol. Vitic. *37*, 206-210.

Ramey, D., A. Bertrand, C. S. Ough, V. L. Singleton and E. Sanders (1986) Effects of skin contact temperature on Chardonnay must and wine composition. Am. J. Enol. Vitic. *37*, 99-106.

Ramey, D. D. and C. S. Ough (1980) Volatile ester hydrolysis or formation during storage of model solutions of wines. J. Agric. Food Chem. *28*, 928-934.

Rankine, B. C. (1984) Hot bottling can be attractive alternative for standard wine. Aust. Grapegrower Winemaker (*252*), 22-23.

Rankine, B. C. (1987) Use of ascorbic and erythorbic acids in winemaking. Aust. Grapegrower Winemaker (*284*), 31.

Ravji, R. J., S. S. Rodriguez and R. J. Thornton (1988) Glycerol production by four common grape molds. Am. J. Enol. Vitic. *39*, 77-82.

Razungles, A. and P. Bidan (1987) Reflexions sur la dégustation: de la necessité d'une standardisation des descripteurs en analyse sensorielle des vins. Rev. Franç. Oenol. *27*(109), 3-10.

Rhein, O. H. and F. Neradt (1979) Tartrate stabilization by the contact process. Am. J. Enol. Vitic. *30*, 265-271.

Ribéreau-Gayon, P. (1977) Oxidative phenomena in grape must. p. 64-70. 3rd Wine Industry Technical Conference. 9-11 Aug. Albany. Australian Wine Research Institute.

Ribéreau-Gayon, P. and A. Lonvaud-Funel (1976) Le gaz carbonique des vins: Son incidence organoleptique. C. R. Acad. Agric. *62*, 491-497.

Robertson, G. L. (1983) Salicylic acid in grapes. Am. J. Enol. Vitic. *34*, 42-43.

Rochard, J., G. Romanet, G. Bureau, and J.-P. Gyr (1988) Essais d'unisation des bouchons de champagne. Rev. Franç. Oenol. *28*(112), 65-66, 68-70, 72-74.

Roufet, M., C. L. Bayonove and R. E. Cordonnier (1987) Étude de la composition lipidique du raisin, *Vitis vinifera* L.: Évolution au cours de la maturation et localisation dans la baie. Vitis *26*, 85-97.

Rozes, N., B. Cuzange, F. Larue and P. Ribéreau-Gayon (1988) Incidence sur la fermentation alcoolique d'une supplémentation du moût de raisin en sulfate d'ammonium. Connaiss. Vigne Vin *22*, 163-167.

Sablayrolles, J. M. and P. Barre (1987) Évaluation des besoins en oxygene de fermentations alcooliques in conditions oenologiques simulées. Rev. Franç. Oenol *27*(107), 34-38.

Sadler, G. D., J. Roberts and J. Cornell (1988) Determination of oxygen solubility in liquid foods using a dissolved oxygen electrode. J. Food Sci. *53*, 1493-1496.

Salgues, M., V. Cheynier, Z. Gunanta and R. Wydle (1986) Oxidation of grape juice 2-S-glutahionyl caffeoyl tartaric acid by *Botrytis cinerea* laccase and characterization of a new substance; 2,5di-S-glutathionyl caffeoyl tartaric acid. J. Food Sci. *51*, 1191-1194.

Salgues, M., C. Dumont and F. Maris (1982) Étude de quelques conditions influencant la filtration des vins sur membrane. Connaiss. Vigne Vin *16*, 257-269.

Saywell, L. G. (1934) Clarification of vinegar. Ind. Eng. Chem. *26*, 981-982.

Scott, R. S., T. G. Anders and H. Hums (1981) Rapid cold stabilization of wine by filtration. Am. J. Enol. Vitic. *32*, 138-143.

Serrano, M. (1988) Filtres: Mesurez leur efficacité. Viti (*126*), 110.

Serrano, M., Ph. Catherinaud and M. Paetzold (1987) Étude comparee de trente-huit diatomées de differentes origines. Connaiss. Vigne Vin. *21*, 279-287.

Simon, R. A., A. Tenscher, C. S. Ough and V. Marinkovich (1988) Adverse reactions to sulfite adducts presented in wine in sulfite sensitive asthmatics. Am. Acad. Allerg. Immun. March 1988, Anaheim.

Simpson, R. F. and G. C. Miller (1983) Aroma composition of aged Riesling wine. Vitis *22*, 51-63.

Simpson, R. F. and G. C. Miller (1984) Aroma composition of Chardonnay wines. Vitis *23*, 143-158.

Singleton, V. L. (1988) *Wine Phenols*. P. 173-218. in "Wine Analysis." eds. H. F. Linsken and J. F. Jackson. Modern Methods of Plant Analysis (New Series) Vol. 6. Springer-Verlag, Berlin.

Singleton, V. L., J. Zaya, E. Trousdale and M. Salques (1984) Caftaric acid in grapes and conversion to a reaction product during processing. Vitis *23*, 113-120.

Somers, T. C. and E. Verrette (1988) *Phenolic composition of natural wine types*. p. 219-257. in "Wine Analysis." eds. H. F. Linksen and J. F. Jackson. Modern Methods of Plant Analysis (New Series) Vol. 6. Springer-Verlag, Berlin.

Somers, T. C. and G. Ziemelis (1985) Flavonol haze in white wines. Vitis *24*, 43-50.

Stackler, B. and C. S. Ough (1979) Differences in the formation of certain acetamides in grape and apple wines. Am. J. Enol. Vitic. *30*, 117-118.

Stafford, P. A. and C. S. Ough (1976) Formation of methanol and ethyl methyl carbonate by dimethyl dicarbonate in wine and model solutions. Am. J. Enol. Vitic. *27*, 7-11.

Strauss, C. R., B. Wilson and P. J. Williams (1985) Taints and off flavours resulting from contamination of wines: A review of some investigations. Aust. Grapegrower Winemaker (*256*), 20, 22, 24.

Sudraud, P. and S. Chauvet (1985) Activité antilevure de l'anhydride sulfureux moléculaire. Connaiss. Vigne Vin *19*, 31-40.

Tanner, H., C. Zanier and H. R. Buser (1981) 2,4,6-Trichloroanisole: Eine dominierende Komponente des Korkgeschmackes. Schweiz. Z. Obst. Weinbau *117*, 97-103.

Thoukis, G. and M. A. Amerine (1956) The fate of copper and iron during fermentation of grape musts. Am. J. Enol. Vitic. *7*, 45-53.

Tredoux, H. G., J. L. F. Kock, P. M. Lategan and H. B. Muller (1987) A rapid identification technique to differentiate between *Saccharomyces cerevisiae* strains and other yeast species in the wine industry. Am. J. Enol Vitic. *38*, 161-164.

Tredoux, H. G., R. P. Tracey and A. Tromp (1986) Killer factor in wine yeast and its effect on fermentations. S. Afr. J. Enol. Vitic. *7*, 105-112.

Unterholzner, O., M. Aurich and K. Platter (1988) Geschmacks- und Gerüchsfehler bei Rotweinen verursacht durch *Schizosaccharomyces pombe* L. Mitt. Klosterneuburg *38*, 66-70.

van Uden, N. (1989a) *Effects of alcohols on the temperature relations of growth and death in yeasts*. p. 77-88. in "Alcohol Toxicity in Yeast and Bacteria." ed. N. van Uden, CRC Press, Inc., Boca Raton, Florida.

van Uden, N. (1989b) *Effects of alcohol on membrane transport in yeasts*. p. 135-146. in "Alcohol Toxicity in Yeast and Bacteria." ed. N. van Uden, CRC Press, Inc., Boca Raton, Florida.

van Vuuren, H. J. J. and L. van der Meer (1987) Fingerprinting of yeasts by protein electrophoresis. Am. J. Enol. Vitic. *38*, 49-53.

van Vuuren, H. J. J. and B. D. Wingfield (1986) Killer yeasts—cause of stuck fermentations in wine cellars. S. Afr. J. Enol. Vitic. 7, 113-118.

Villettaz, J. C. (1988) Les colloides du moût et du vin. Rev. Franç. Oenol. *28*(111), *23*, 25-27.

Villettaz, J. C., D. Steiner and H. Trogus (1984) The use of a beta glucanase as an enzyme in wine clarification and filtration. Am. J. Enol. Vitic. *35*, 253-256.

Wagener, W. W. D., C. S. Ough and M. A. Amerine (1971) The fate of some organic acids added to grape juice prior to fermentation. Am. J. Enol. Vitic. *22*, 167-171.

Watts, D. A., C. S. Ough and W. D. Brown (1981) Residual amounts of proteinaceous additives in table wines. J. Food Sci. *46*, 681-683, 687.

Weiller, H. G. and F. Radler (1976) Über den Aminosäurestoffwechsel von Milchsäurebacterien aus Wein. Z. Lebensm. Unters. Forsch. *161*, 259-266.

White, B. B. and C. S. Ough (1973) Oxygen uptake studies on grapes. Am. J. Enol. Vitic. *24*, 148-152.

Williams, J. T., C. S. Ough, and H. W. Berg (1978) White wine composition and quality influenced by method of must clarification. Am. J. Enol. Vitic. *24*, 92-96.

Williams, L. A. and R. Boulton (1983) Modeling and prediction of evaporative ethanol loss during wine fermentation. Am. J. Enol. Vitic. *34*, 234-242.

Williams, P. J., C. R. Strauss, A. P. Aryan and B. Wilson (1987) Grape flavour—a review of some pre- and post-harvest influences. p. 111-116, ed. T. H. Lee, Proc. 6th Aust. Wine Ind. Tech. Conf. Adelaide, 1986.

Wilson, D. L. (1985) Storage of wine using inert gas for prevention of oxidation. Aust. Grapegrower Winemaker (*256*), 122-124, 126-127.

Wise, A. V. and R. M. Pool (1989) Wines and winegrapes: A success story for Long Island agriculture. New York Food and Life Sci. Quart. *19*(1), 25-27.

Wucherpfennig, K., H. Dietrich, and K. Otto (1987) Neuentwicklungen zur Stabilisierung von Wein gegen kristallin Ausscheidungen. Wein-Wissen. *43*, 241-265.

Wucherpfennig, K., K. Otto, and S. Nebel (1988) Über den Einfluss von Calciumionen auf die Auskristallisation von Weinstein. Wein-Wissen. *43*, 339-349.

Zironi, R., N. Frega, L. S. Conte and G. Lercker (1984) Éffets de la formation de la pourriture acidé sur la composition de la fraction lipidique des différentes parties du grain de *Vitis vinifera* cv. Fortana. Vitis *23*, 93-105.

Bibliography

Amerine, M. A., H. W. Berg, R. E. Kunkee, C. S. Ough, V. L. Singleton and A. D. Webb (1980) *Technology of Winemaking*. 4th ed. 794 p. Avi Publishing Co., Westport, Connecticut.

Amerine, M. A., and E. B. Roessler (1983) *Wines, Their Sensory Evaluation*. 2nd ed. 432 p. W. H. Freeman and Co., San Francisco.

Amerine, M. A. and A. J. Winkler (1963) *California Wine Grapes: Composition and Quality of Their Musts and Wines*. Calif. Expt. Sta. Bull. 794, 83 p. University of California.

Champagnol, F. (1984) *Elements de Physiologie de la Vigne et de Viticulture Generale*. 351 p. S.A.R.L., Montpellier.

Dittrich, H. H. (1987) *Mikrobiologie des Weines*. 2nd ed. 375 p. Eugen Ulmer GmbH and Co., Stuttgart, Germany.

Farkaš, J. (1988) *Technology and Biochemistry of Wine*. Vols. 1 and 2. 744 p. Gordon and Breach Science Publishes, S.A., Switzerland.

Hallgarten, F. (1986) *Wine Scandal*. 183 p. G. Weidenfeld and Nicolson, Ltd., London.

Kreger-van Rij, N. J. W. (1984) *The Yeasts, a Taxonomic Study*. 3rd ed. 1082 p. Elsevier Science Publishers. B. V., Amsterdam.

Linskens, H. F. and J. F. Jackson (1988) *Wine Analysis*. 381 p. "Modern Methods of Plant Analysis." (New Series) Vol. 6, Springer-Verlag, Berlin.

Nagy, S., J. A. Attaway and M. E. Rhodes (1988) *Adulterations of Fruit Juice Beverages*. 563 p. Marcel Dekker, Inc., New York.

Nykanen, L. and P. Lehtonen (1984) *Flavour Research of Alcoholic Beverages Instrumental and Sensory Analysis*. 335 p. Foundation for Biochem. Indust. Ferment. Res. Vol. 3., Helsinki.

Ouchi, G. I. (1987) *Personal Computers for Scientists. A Byte at a Time*. 276 p. Amer. Chem. Soc., Washington, D. C.

Ough, C. S. (1988) *Determination of sulfur dioxide in grapes and wines.* In "Modern Methods of Plant Analyses." Wine Analysis. Eds. H. L. Linskens and J. F. Jackson. Vol 6, p. 339-358, Springer-Verlag, Berlin.

Ough, C. S. and M. A. Amerine (1966) *Effects of Temperatures on Winemaking*. Calif. Agric. Expt. Sta. Bull. 827, 1-36, University of California.

Ough, C. S. and M. A. Amerine (1988) *Methods for Analysis of Musts and Wines*. 377 p. John Wiley and Sons, Inc., New York.

Paronetto, L. and L. Paronetto (1986) *Ausiliari Chimici e Biologici in Enologia*. 941 p. INTEC Editrice, Verona.

Peynaud, E. (1984) *Knowing and Making Wine*. 391 p. John Wiley and Sons, Inc., New York.

Peynaud E. (1987) *The Taste of Wine: The Art and Science of Wine Appreciation*. 258 p. Macdonal Orbis, London.

Rankine, B. C. (1989) *Making Good Wine. A Manual of Winemaking Practice for Australia and New Zealand*. 374 p. The Macmillan Co. of Australia, Pty.

Ribéreau-Gayon, J., E. Peynaud, and P. Sudraud (1975). *Traite d'Oenologie. Science et Techniques du Vin II. Caracteres des Vins. Maturation du Raisin, Levures et Bacteries*. 556 p. Dunod, Paris.

Rose, A. H. and J. S. Harrison (1987) *The Yeasts*. 2nd ed. Vol. 1 "Biology of yeasts." 423 p. Academic Press, Inc., London.

Stone, H. and I. L. Sidel (1985) *Sensory Evaluation Practices*. 311 p. Academic Press, Inc., New York.

Wedzicha, B. L. (1984) *Chemistry of Sulfur Dioxide in Foods*. 381 p. Elsevier Applied Science Publishers, New York.

Winkler, A. J., J. A. Cook, W. M. Kliewer and L. A. Lider (1974) *General Viticulture*. 710 p. University of California Press, Berkeley.

Appendixes

Appendix I. Amount (ml) of 5% w/v of SO_2 to Achieve Indicated Amounts

Volume of wine[a] (gals)	Desired SO_2 concentration (mg/L)					
	25	50	75	100	125	150
1/2	.95	1.9	2.85	3.8	4.75	5.7
1	1.9	3.8	5.7	7.6	9.5	11.4
2	3.8	7.6	11.4	15.2	19.0	22.8
3	5.7	11.4	17.1	22.8	28.5	34.2
4	7.6	15.2	22.3	30.4	38.0	45.6
5	9.5	19.0	28.5	38.0	47.5	57.0
6	11.4	22.8	34.2	45.6	57.6	68.4
7	13.3	26.6	39.9	53.2	66.5	79.8
8	15.2	30.4	45.6	60.8	76.0	91.2
9	17.6	35.2	51.3	68.5	85.5	102.6
10	19.0	38.0	57.0	76.0	95.0	114.0
12	22.8	45.6	68.4	91.2	114.0	137.0
14	26.6	53.2	79.8	106.4	133.0	159.6
16	30.4	60.8	91.2	121.6	152.0	182.4
18	35.2	70.4	102.6	137.0	171.0	205.2
20	38.0	76.0	114.0	152.0	190.0	228.0
30	57.0	114.0	171.0	228.0	285.0	342.0
40	76.0	152.0	228.0	304.0	380.0	456.0
50	95.0	190.0	285.0	386.0	475.0	570.0
100	190.	380.0	570.0	772.0	950.0	1140.0

[a]For larger volumes multiply by appropriate factor.

Appendix II. Interconversion of Metric and American Units

Length

1 centimeter	=	0.3937 inches
1 inch	=	2.54 centimeters
1 meter	=	3.28 feet
1 foot	=	0.3049 meters

Area

1 square meter	=	1.196 square yards
1 square yard	=	0.836 square meters (centares)
1 hectare	=	2.471 acres
1 acre	=	0.405 hectare

Volume

1 liter	=	0.2642 gallon
1 gallon	=	3.7853 liters
1 hectoliter	=	26.42 gallons
1 gallon	=	0.03785 hectoliters
1 ml	=	0.0338 fluid ounces
1 fluid ounce	=	29.6 ml

Weight

1 gram	=	0.0353 ounces or 0.0022 pound
1 pound	=	454 grams
1 kilogram	=	2.20 pounds

Weight/volume

1 mg/liter	=	0.083 lb/1000 gal
1 g/hectoliter	=	0.83 lb/1000 gal

Appendix III. Refractometer Correction for °Brix

Temperature °C	°Brix [Sucrose (g/100 g solution)] 0	5	10	15	20	25	30
				Subtract			
10	0.50	0.54	0.58	0.61	0.64	0.66	0.68
11	0.46	0.49	0.53	0.55	0.58	0.60	0.62
12	0.42	0.45	0.48	0.50	0.52	0.54	0.56
13	0.37	0.40	0.42	0.44	0.46	0.48	0.49
14	0.33	0.35	0.37	0.39	0.40	0.41	0.42
15	0.27	0.29	0.31	0.33	0.34	0.34	0.35
16	0.22	0.24	0.25	0.26	0.27	0.28	0.28
17	0.17	0.18	0.19	0.20	0.21	0.21	0.21
18	0.12	0.13	0.13	0.14	0.14	0.14	0.14
19	0.06	0.06	0.06	0.07	0.07	0.07	0.07
				Add			
21	0.06	0.07	0.07	0.07	0.07	0.08	0.08
22	0.13	0.13	0.14	0.14	0.15	0.15	0.15
23	0.19	0.20	0.21	0.22	0.22	0.23	0.23
24	0.26	0.27	0.28	0.29	0.30	0.30	0.31
25	0.33	0.35	0.36	0.37	0.38	0.38	0.39
26	0.40	0.42	0.43	0.44	0.45	0.46	0.47
27	0.48	0.50	0.52	0.53	0.54	0.55	0.55
28	0.56	0.57	0.60	0.61	0.62	0.63	0.63
29	0.64	0.66	0.68	0.69	0.71	0.72	0.72
30	0.72	0.74	0.77	0.78	0.79	0.80	0.80

For correction above 30°Brix see Ough and Amerine (1989).

Appendix IV. Temperature Correction for Brix Hydrometers Calibrated at 20°C.

Temperature	Observed sugar (°Brix)						
°C	0	5	10	15	20	25	30
				Subtract			
15	0.20	0.22	0.24	0.26	0.28	0.30	0.32
16	0.17	0.18	0.20	0.22	0.23	0.25	0.26
17	0.13	0.14	0.15	0.16	0.18	0.19	0.20
18	0.09	0.10	0.10	0.11	0.12	0.13	0.13
19	0.05	0.05	0.05	0.06	0.06	0.06	0.07
				Add			
21	0.04	0.05	0.06	0.06	0.06	0.07	0.07
22	0.10	0.10	0.11	0.12	0.12	0.13	0.14
23	0.16	0.16	0.17	0.17	0.19	0.20	0.21
24	0.21	0.22	0.23	0.24	0.26	0.27	0.28
25	0.27	0.28	0.30	0.31	0.32	0.34	0.35
26	0.33	0.34	0.36	0.37	0.40	0.40	0.42
27	0.40	0.41	0.42	0.44	0.46	0.48	0.50
28	0.46	0.47	0.49	0.51	0.54	0.56	0.58
29	0.54	0.55	0.56	0.59	0.61	0.63	0.66
30	0.61	0.62	0.63	0.66	0.68	0.71	0.73
35	0.99	1.01	1.02	1.06	1.10	1.13	1.16

Appendix V. Brix, Baumé and Specific Gravity Equivalents.

°Brix (Balling)	Baumé[a]	Specific gravity at 20°/20°C	°Brix (Balling)	Baumé[a]	Specific gravity at 20°/20°C
0.0	0.00	1.00000	21.2	11.8	1.08823
0.2	0.1	1.00078	21.4	11.9	1.08913
0.4	0.2	1.00155	21.6	12.0	1.09003
0.6	0.3	1.00233	21.8	12.1	1.09093
0.8	0.45	1.00311	22.0	12.2	1.09183
1.0	0.55	1.00389	22.2	12.3	1.09273
2.0	1.1	1.00779	22.4	12.45	1.09364
3.0	1.7	1.01172	22.6	12.55	1.09454
4.0	2.2	1.01567	22.8	12.7	1.09545
5.0	2.8	1.01965	23.0	12.8	1.09636
6.0	3.3	1.02366	23.2	12.9	1.09727
7.0	3.9	1.02770	23.4	13.0	1.09818
8.0	4.4	1.03176	23.6	13.1	1.09909
9.0	5.0	1.03586	23.8	13.2	1.10000
10.0	5.6	1.03998	24.0	13.3	1.10092
11.0	6.1	1.04413	24.2	13.45	1.10193
12.0	6.7	1.04831	24.4	13.55	1.10275
13.0	7.2	1.05252	24.6	13.7	1.10367
14.0	7.8	1.05667	24.8	13.8	1.10459
15.0	8.3	1.06104	25.0	13.9	1.10551
16.0	8.9	1.06534	25.2	14.0	1.10643
17.0	9.4	1.06968	25.4	14.1	1.10736
17.4	9.7	1.07142	25.6	14.2	1.10828
18.0	10.0	1.07404	25.8	14.3	1.10921
18.4	10.2	1.07580	26.0	14.45	1.11014
19.0	10.55	1.07844	26.2	14.55	1.11106

Appendix V (continued)

°Brix (Balling)	Baumé[a]	Specific gravity at 20°/20°C	°Brix (Balling)	Baumé[a]	Specific gravity at 20°/20°C
19.2	10.65	1.07932	26.4	14.65	1.11200
19.4	10.8	1.08021	26.6	14.8	1.11293
19.6	10.9	1.08110	26.8	14.9	1.11386
19.8	11.0	1.08198	26.8	14.9	1.11386
20.0	11.1	1.08287	27.0	15.0	1.11480
20.2	11.2	1.08376	27.2	15.1	1.11573
20.4	11.35	1.08465	27.4	15.2	1.11667
20.6	11.45	1.08554	27.6	15.3	1.11761
20.8	11.55	1.08644	27.8	15.45	1.11855
21.0	11.7	1.08733	30.0	16.57	1.12898

[a]Approximated by the use of this equation: °Brix = 1.8 × Baumé. These values check against those of Table 109 of Circular C440 of the National Bureau of Standards, Polarimetry Saccharimetry, and the Sugars. Reducing sugar is approximately 2.0 less than the °Brix. The °Brix (Balling) times 0.52 will give the approximate prospective alcohol. (See Ough and Amerine, 1989, for more information.)

Index

Acetaldehyde
  flor sherry, 198
  malolactic bacteria inhibitions, 190
Acetic acid
  flor sherry, 198
Acetoin, 128,133,137
  flor sherry, 198
Acetyl-tetrahydropyridine, 188
Acidity
  additions, 83
  sensory effects, 247
Acids
  metal pickup, 180
Active amyl alcohol, 131
Additives
  adjuncts to change flavor or composition, 285-287
  contaminants, 274-275
  fraud detection, 288-290
  illegal yeast inhibitors, 281-284
  legal for grapes and wines, 275-281
  natural yeast inhibitors, 287-288
Aeration-oxidation method, 270
Agglomerate corks, 210
Aging
  fermentation esters, 216
  fining results, evaluation of, 221-222
  malolactic effects of, 217
  red wines, 220-225
  rosé or blush wines, 219-220
  sulfur dioxide, 219
  wood, effects of, 216
  yeast contact, 217-218
AGRICOLA, 271
Aktuan, 190
Allergies, 153
Aluminum, 181
Amelioration, 83
Amino acids
  assimilation by malolactic bacteria, 189
  detection of fraud, in juices, 290
  skin contact, 107
  vineyard fertilization effects, 89
  yeast assimilation, 106,107
Amino acids, synthesis
  active aldehyde or pyruvate condensation, 131
  families of, 132
  schemes, 130
Ammonia, 86
Analysis of variance, 255
Anions, 143-144
Anthocyanin glucosides, 142
Arabitol, 86
Arginase, 132
Aromas
  botrytis, 246
  esters, 246
  off, 246,247
Aroma wheel, 254-255
Arsenic, 102
Ascorbic acid
  juice, amounts in, 85
  oxygen, reactions with, 280-281
  sulfur dioxide, use with, 281

Bag-in-the-box, 232,234
Barbera, 36
Barrels
  compositional and sensory changes, wines, 220,225,227

handling and holding, 217,220-221
making and preserving, 225-227
Behenic acid, 90
Bentonite
metal contamination, 148
protein stabilization, 179
time of addition, 180
use and preparation, 146,147-148
Bentotest, 179
Bidule, 206
Bitterness, 248
Blanketing, inert gas, 230
Blending formula, 196
Blue casse (ferric tannate), 181
Blue fining, 182
Blush wines, 58,126
Botrytis cinerea, 38,43,85
β-Glucans, 164
laccase, 61
sensory aroma, 246
sugar alcohols, effect on, 86
Botrytis wine, 119-120
Bottle bouquet, 245
Bottling
closures, home winery, 299
oxygen, elimination of, 232
quality control, 231
Brettanomyces, 184-185
Brimstone, 48
Brix, 16
Brix/acid, 19
Brix × pH, 19
fermentation, measurement during, 260
method of analysis, 259-261
Bromide, 144
Browning, 169-170
2,3-Butanediols, 128,133,137
2-Butanol, 188
s-Butyl methoxypyrazine, 30

Cabernet franc, 29
Cabernet Sauvignon, 29,59,245
Cadmium, 143
Caffeic acid, 61,65
Caffeol tartaric acid (caftaric acid), 79
Calcium
juice, amount in, 91
stabilization, 175
Captan, 101,102
Carbonated wines, 212-214
Carbon dioxide
counter pressure, Charmat process, 211
desired amount in table wines, 230-231
Carignane, 29
Carmine, 30
Casein, 149
preparation of, 170-171
Catawba, 202
Catechins, 61,142
Cavitator, 198,200
Centrifuges, 154
Centurian, 29
Champagne (sparkling wine)
amino acid comparisons, 212
base material, 58,59,60
Charmat, 211
clarification and bottling, 206-209
continuous process, 211-212
fermentation, secondary, 203-205
Chardonnay, 35,201,245
composition, amino acids, 88
skin contact, 119
Charmat (bulk or tank process), 210-212
Chemical abstracts, 271-272
Chemical analysis
components, legally defined, 258
routine for juice and wine, 258-270
water, 270-271
Chenin blanc, 30,202
Chloride, 91
Chlorine, 54
bad corks, 238
Chlorogenic acid, 61

Citric acid, 83,144
  amounts in juice, 85
  iron chelating properties, 181
Citric acid cycle, 133
Citronellol, 38
Cleaning
  equipment and tanks, 52
  home winery, 297
Climatic effects, 6-8,23,28
Cold stabilization, 154
Colloids, 61
Colombard (see French Colombard)
Color, 33,59
  anthocyanins, 176
  fining, 153
  sensory aspects, 244
  stability and effects of pH and sulfur dioxide, 176
  thermovinification, 60
Color extraction
  methods, 120-121
Composition
  barrel aging changes, 227
  reasons to measure, 137
Computer searches, 271
Concentrate
  for wine use, 46-47
Contact tartrate stabilization method, 172-173
Contaminants, 273,274-275
Copper, 91
  amounts, 142,143
  hydrogen sulfide removal, 177
  stability, 182-183
Corks
  champagne, 209-210
  corkiness, causes, 238-239
  paraffin coating, 239
p-Courmaroyl tartaric acid, 79
Crusher-stemmers, 56
  fruit condition, 56,58

β-Damacenone, 64
Decanters, 77
Delaware, 202
Depth filter, 156
Descriptive terminology, 252-254
Dextrin, 85
Diacetyl, 128,133
DIALOG, 271
Diammonium phosphate (DAP), 105,107
  hydrogen sulfide, use for, 177
Diatomaceous earth, 160
Diethylene glycol, 285
Dimethyl dicarbonate
  use and properties, 278-280
Diendiol-1, 38
Dosage, 209
Downy mildew, 7
Drainers, 65,74
Dry yeast
  use and properties, 96,97,99
Durif, 36

Ebulliometer, 267
Egg whites
  amount and preparation, 151
Ehrlich-Neubauer and Fromberg scheme, 130
Epoxy coatings, 49
Erythorbic acid (see ascorbic acid)
Erythritol, 86
Esterase, 114,188
Esters, 138
  botrytis, effect on, 43
  chemistry of, 113-115
  sensory aspects, 246
Ethanol
  chemical-sensory effects, 248
  federal tax rate, 137-138
  fermentation losses during, 108
  measurement in wine, 267
Ethyl carbamate, 133
Ethylenediamine tetraacetate (EDTA), 52
Ethyl mercaptan, 178
Ethyl sorbate, 277
Ethyl tartrate esters, 114,115
2-Ethoxyhex-3,5-diene, 277

Euparen, 102
  inhibition of malolactic bacteria, 190

Fatty acids
  in grapes, 90
  ghost cells, 102
  malolactic bacteria, inhibition, 192
  yeast, inhibition, 102
Fermentation
  brix, changes during, 117
  dryness checks, 118
  factors affecting, 108
  rate of, models, 108,113
  traps, 169
Fermentation bouquet, 113
Feruloyl tartaric acid, 79
Filter aid (diatomaceous earth, kieselguhr)
  composition, 159
  use, 160
Filters
  commercial types and use, 154-157
  home use and types, 166-167
Filtration
  choice of, 160-166
  prediction models for, 166
  sterile, for bottling, 231
Fining
  gaining expertise, 153,154
  materials, for red wine, 150-151
  materials, for white wines, 146-151
  red wines, in barrels, 220
  trials, 149
Flor sherry
  blending, 196,197
  submerged culture, 198-201
  tank or barrel fermentation, 197
Fluoride, 144
Food and Technology Abstracts, 271
Foils
  lead usage, 240
French Colombard, 202
French hybrids, 23,28,84
Fructose, 85
Fruit quality, 42-43
Fumaric acid, 83,190
Fungicides
  clarification additives, effect on, 103
  legal importance, 103
  yeast inhibition, 99,101

Gallic acid, 65
Gelatin
  bloom number, 151
  preparation, 151
Gelatin/silica gel
  removal of tannin or protein, 148
Geosmin, 237
Geraniol, 38
Gewürztraminer, 36,38,245
  skin contact, 119
Ghost cells, 102,192
Glacturonic acid
  juice, amount in, 86
β-Glucans, 164,166
Gluconic acid
  juice, amount in, 86
β-Gluconase, 62
Glucose, 85
Glucose/fructose, 119
Glutathione, 79
2,5-S-glutathionyl caffeoltartaric acid, 79
Glycerol, 127
Glycolysis
  pathway, 127
  temperature control, 128
Glycoproteins, 208
Glycosidases, 36
Glycosides
  anthocyanin, 176
  terpene, 36
Grape
  cultivars (varieties), 23-29
  growing conditions, 6-8

harvesting, 39-42
maturity, 16-22
pests, 5,8
sampling (see vineyard sampling)
temperature control, after harvest, 49,63
viticultural characteristics, 23
Grape concentrate, 84
Grape juice
composition, 85-91
sensory evaluation, 252-253
Grenache, 59,245
Gums, 85
Gushing, 240-242

Hard water, 54
Haze, sensory significance, 244
Headspace, 229,232
Hexanol, 89
Higher alcohols (fusel oils), 130
Histamine, 189
Home winery, 291-296
Hot bottling, 232
Hubach test, cyanide, 182
Hydrogen peroxide
ascorbic acid, formation from, 281
Hydrogen sulfide, 128-130
causes, 116-117,177
treatments for, 117,177-178
yeast, effect of, 130
p-Hydroxycinnamic acid, 65
Hydroxymethylfurfural, 138

Immobilized yeast, 204
Information retrieval, 271-272
Ingraham-Guymon-Crowell scheme, 130
Insoluble solids, fermentation, 82
Iron, 91,142-143
Iron stability
safe levels, 182
tests for, 181
Irrigation, 11

Isinglass
amount and method, 150
Isoamyl alcohol, 131
Isobutyl alcohol, 131
Isopropyl methoxypyrazine, 30

Juice
contact time and settling, 67
heat treatments, 64
home winemaker, separation and sulfur dioxide, 71,74
storage, 48
yields, 6,77,79

Kampferol, 183
α-Ketoglutaric acid, 131
Killer yeast, 99

Labels, 211,240
Laboratory, home winery, 299
Labrusca, 84
Laccase, 61
Lactic acid, 144
Lactic acid bacteria, 186-187
Lactobacillus sp., 186
Lead, 143,240
Leuconostoc sp., 187
Linalool, 38
Linoleic acid, 90
Linolenic acid, 90
Lipids, 90,145

Maceration carbonique, 52,124-126
Magnesium, 91
Malbec, 29
Malic acid, 83,144
amount in, juices, 85
fraud detection, juices, 290
Malolactic fermentation, 66
growth and fermentation, 186-193
measurement, 193-194
red wines, 220

sparkling wines, 202
starters, 192
Mannitol, 86
Masking, 176
Membrane filters, 156-157,162
cleaning, 164
Membrane press, 65
Merlot, 29
Meso-inositol, 86
Metals, 180-183
Metals and other cations, 142-143
Methanol, 89,286-287
poisoning, 286-287
Methionine, 130
2-Methoxy-3-isobutylpyrazine, 29-30
Methyl anthranilate, 29
2-Methyl isoborneol, 238
Meunier (see Pinot Meunier)
Microbiological
contamination, 237
stability, 183-185
MOG (material other than grape), 42
Mold and rot
irrigation, effects on, 7,11
measurement of, 42
Molecular sulfur dioxide, 100
Monoamine oxidase inhibitors, 189
Monobromo- and monochloroacetic acids, 282,285
Muscadine, 84
Muscat blanc, 36,38
Muscat of Alexandria, 38,245
Muscat, varietal aroma
sensory aspects, 245

Napa Gamay, 59
Nerol, 38
Nitrate, in juice, 86
Nitrioltriacetate, 52
5-Nitrofuryl acrylic acid, 283
Nitrogen
compounds, amounts of, 139
stripping, gas, 229
total, effects of bentonite, 139
Noble rot (botrytis), 84
Non-flanonoid phenols, 142,225

Oak shavings
ellagic acid deposits, 217-218
pressing, use during, 227-228
Off-flavors, 56,58,249
Oidium (see powdery mildew)
Oleic acid, 90
Orange muscat, 36,38,245
Organic acids, 144
Organic sulfides, 178
Orthene-50, 102
Oxidation
aging, white wines, 219
ascorbic acid, 280-281
bottling, elimination during, 232
color loss, 228
filtration, problems during, 166
hydrogen sulfide, 178
inert gas use, 230
polyethylene, 46
polyphenol oxidase, 168
protection from, 146-147
white and rosé, 169
wines, oxygen uptake, 221,228-229
Oxygen
enzymatic action on, in juice, 61
removal, 229-230
yeast viability, effect on, 97

Paired tests, 255
Palmitic acid, 90
Pantothenate, 130
Pasteurization, protein stabilization, 179
Pectic enzymes, 48,114
yield and clarification, 79-81
Pectins, 62,85
Pediococcus sp., 187
Pentosans, 85
Perlites, composition, 159
Peronospora (see downy mildew)

Petite Syrah, 35
pH, 66
  "extent of change" equation, 67
  measurement, 262,264
  red pigments, effect on, 176
Phenols, 61
  amounts, 60,141,142
  barrels, kinds of, 226
  fining, removal by, 151
  measurements, methods, variation in, 89
  sensory causes and effects, 248
  stability treatment, PVPP, 150
  types of, 89
Phosphate, 91,144
Phospholipids, 90
Pimaricin, 285-286
Pinking, 169
Pinot Meunier, 34,201
  amino acid composition, 88
Pinot noir, 30,201
  amino acid composition, 88
  clonal trials, 31-32
  eucalyptus leaves, 124
  stem contact, 59,121,124
Plate and frame filters, 160-161
Polyethylene closure, champagne, 211
Polyphenol oxidase (PPO), 62,79
  tyrosinase, 61
Polysaccharides, 166
Polyvinylpolypyrrolidone (PVPP), 77,170
  phenol removal, amounts and method, 150
Potassium, 91
Powdery mildew, 7
Presses, types, 75,77
Pressing
  degree of, 60,62-63,74
  raisins or shrivel, effect of, 74
Principal component analysis, 255
Proline/arginine, 86
Propanediols, 133
n-Propanol, 131
Propionyltetrahydropyridine, 188
Propylene glycol, 285-286
Protchatechic acid, 61
Proteases, 188
Protein, 62
  juice, amount in, 86
  stability, 178-180
  ultra filtration, removal by, 179-180
Pruning and trellises, 12
Pumps, for crushed grapes, 52
Pyrazines, 245
Pyrocatechol, 61

Quercetin, 183
o-Quinone reactions, 79

Racking
  red wines, oxidation, 167,169
  time, 118
Racks for barrels, 221
Raisins, for wine, 46
Rectified concentrate, 47-48
Reducing sugar, Dextro Check, 261
Refractometer, 17-18,260
Refrigeration, 52
Riddling, 206-207,209
Ripper method, 269
Robotics, 206
Rootstock, 5,8
Ropiness, 245
Rosé
  process and varieties, 59
  treatment after press, 126
Rubired, 46
Ruby Cabernet, 29

Salicylic acid, 283
Salvador, 46
Sauvignon blanc, 30,245
Sauternes, 119
Shizosaccharomyces sp., 185
Scion, 5

Semillon, 30,245
Sensory examination
  fining experiments, 221,225
  statistical evaluation, 255
  systems for, use and choice, 250-252
Settling, 67
Shikimic acid pathway, 126
Shiraz, 35
Skin contact
  methanol, 89
  red grapes, factors and changes, 121-124
  white grapes, factors and changes, 64-65,119
Sodium, 91
Sodium azide, 282,284
Sodium triphosphate, 52
Soil treatment, 10
Solids in juice
  effect on fusel oils, 131
Sorbic acid, 277-278
Sorbitol, fraud detection, 290
Sparkloid, 149
Spider diagrams, 255,256
Spoilage organisms, barrels, 220-221
Stainless steel, 49-50
  chlorine, effect on, 54
Stearic acid, 90
Stem contact, 56,59,212
"Sterile", 283
Sterilizing
  bottling, 236-237
  filters, 157
  home winery, 297
Sterols, 90-91
  yeast viability, 97
Storage
  bins, home winery, 297-298
  temperature, 242-243,299
Stuck fermentations, 116
Student-T test, 255
Style, 79
Submerged culture, flor sherry, 198-201
Succinic acid, 128
Sucrose, 204
Sugar
  additions, 83
  fraud detection, 288-289
  pathways variation, 288-289
  sensory effect, 248
Sulfur dioxide, 48,59,67-68
  aging and storage, 219
  amounts, 71
  color retention, 219
  dusting, hydrogen sulfide formation, 177
  Henderson-Hasselbalch equation, 100
  laccase, 61
  malolactic inhibition, 190,192,220
  measurement, 269-270
  medical and legal aspects, 275-277
  metabisulfite, 70
  oxidation, 79,242
  oxygen, reaction with, 229
  red pigments, reaction with, 176
  stability, microbiological, 183
  yeast, effect on, 94,96
Sulfate, 91,144
Sulfate/sulfite/sulfide pathway, 128-130
Sur lie, 216-217
Syrah, 35

Table wine, freshness, 230-231
Taint, 215
Tangential filter, 164
Tartaric acid, 83,144
  amount in juice, 85
  fraud detection, 290
Tartrate stabilization, 46,171-176
  addition of D(-), D,L tartrate, 176
Taste, 246-248
Temperature, fermentation, 106,108,205

α-Terpeneol, 38
Terpenes, 36-37,43,115
Thermalvinification, 60-61, 64
α-, β-Thujone, 287
Tinta Portuguese cultivars, 246
Tokay, 119,138
Total titratable acidity, 262-264
Trans-Caffeic acid, 65
Transfer process, 209-210
Trans-Furan linalool, 38
Trans-Geranic acid, 38
Transport, temperature of, 242
Trehalose, 86
Trichloroanisoles, 237-238
Tyramine, 189
Tyrosinase, 61

Urea, 132-133
  amounts, 144
  metabolism, malolactic bacteria, 188-189
Ultra filtration, 164,179-180

Vanillin, 226
Vine planting, 11
Vineyard, 5-16
  sampling, 18-19
Vitamins, 89,105
Vitis lubruscana, 23,28
Vitis rotundifolia, 23
Vitis sp., glycoside differences, 176
Vitis vinifera, cultivars not recommended, 39
Volatile acidity, 264,267

Water tests, 270
White casse (ferric phosphate), 181
White Riesling, 36,38,245,246
  aging, odor, 219
  noble rot, 84
Wine coolers, 288
Wine quality, 33-34
Wine taster, 251-252
Wine tasting, 256-257
Wood borer, 226
Woodruff, 287

Yeast
  biomass, 128
  contamination, 237
  culture preparation and choice of, 94-96
  growth factors, 103-107
  identification, DNA, 93
  inhibitors, 232-236
  malolactic bacteria, effect on, 191
  metal adsorbtion, 181
  pressure effects on, 211
  strains, 109-110
  sulfur dioxide, acclimatization to, 202
Yields, 12,16

Zinc, 142
Zinfandel, 36,40,59,245,246

WITHDRAWN
Return To
Pictou Campus
Nova Scotia Community College
Resource Centre